# Drought Monitoring, Prediction and Impacts

# Drought Monitoring, Prediction and Impacts

Guest Editor

**Muhammad Abrar Faiz**

MDPI

Basel • Beijing • Wuhan • Barcelona • Belgrade • Novi Sad • Cluj • Manchester

*Guest Editor*
Muhammad Abrar Faiz
School of Water Conservancy
and Civil Engineering
Northeast Agricultural
University
Harbin
China

*Editorial Office*
MDPI AG
Grosspeteranlage 5
4052 Basel, Switzerland

This is a reprint of the Special Issue, published open access by the journal *Atmosphere* (ISSN 2073-4433), freely accessible at: https://www.mdpi.com/journal/atmosphere/special_issues/HBYJEKG0MC.

For citation purposes, cite each article independently as indicated on the article page online and as indicated below:

| |
|---|
| Lastname, A.A.; Lastname, B.B. Article Title. *Journal Name* **Year**, *Volume Number*, Page Range. |

**ISBN 978-3-7258-5501-8 (Hbk)**
**ISBN 978-3-7258-5502-5 (PDF)**
**https://doi.org/10.3390/books978-3-7258-5502-5**

# Contents

# About the Editor

**Muhammad Abrar Faiz**

Muhammad Abrar Faiz is a professor and doctoral supervisor at the School of Water Conservancy and Civil Engineering, Northeast Agricultural University, Harbin, China. His research expertise includes hydrological system analysis and water resource management, with a strong emphasis on drought monitoring and assessment. He has led the National Natural Science Foundation of China's Foreign Young Scholar Research Fund project on agricultural drought development, recovery, and propagation. He has contributed to several major national and provincial research initiatives focused on hydrological resilience and water resource management. Prof. Faiz has authored over fifty peer-reviewed articles, including first-authored contributions to high-impact journals such as the *Journal of Hydrology*, *Agricultural Water Management*, *International Journal of Digital Earth*, *Journal of Cleaner Production*, and *Atmospheric Research*. His recent work, published in the *Journal of Advanced Research* (2025), examines interactions among soil moisture, vegetation health, and drought severity, offering new insights into ecosystem resilience under extreme climate conditions. Recognized as a "Northeast Agricultural Scholar – Core Talent A" at his university, he has also served as a peer reviewer for leading international journals in hydrology, environmental science, and climatology. Through his research on drought dynamics and climate–water interactions, he provides valuable expertise as the Guest Editor of this Special Issue of Atmosphere.

# Preface

This reprint compiles recent scientific advances in monitoring, simulating, and predicting drought amid the accelerating impacts of climate change. Its scope includes atmospheric processes, hydrological responses, agricultural effects, and socio-ecological consequences, emphasizing the complex interactions that influence drought risk. The goal is to create a comprehensive platform that integrates diverse methods, from remote sensing and climate modeling to data fusion and risk assessment, to improve the understanding and management of drought hazards.

The reason for compiling this reprint is the urgent need to address the increasing challenges caused by more frequent and severe droughts. These events threaten food security, strain water supplies, and weaken ecosystem resilience, making it crucial to develop innovative monitoring tools and adaptive strategies.

This reprint is aimed at researchers, practitioners, and policymakers involved in atmospheric sciences, hydrology, agriculture, and environmental management. By providing interdisciplinary insights, it aims to support informed decision-making and encourage collaboration to build resilience against one of the most urgent climate-related issues of our time.

**Muhammad Abrar Faiz**
*Guest Editor*

*atmosphere*

MDPI

*Article*

# Deep Learning for Flash Drought Detection: A Case Study in Northeastern Brazil

**Humberto A. Barbosa [1,*], Catarina O. Buriti [2] and T. V. Lakshmi Kumar [3]**

[1] Laboratório de Análise e Processamento de Imagens de Satélites (LAPIS), Instituto de Ciências Atmosféricas, A. C. Simões Campus, Universidade Federal de Alagoas, Maceió 57072-900, Brazil
[2] National Semi-Arid Institute (INSA), Ministry of Science, Technology, and Innovations (MCTI), Campina Grande 58100-000, Brazil
[3] School of Environmental Sciences, Jawaharlal Nehru University, New Mehrauli Road, New Delhi 110 067, India; lakshmikumar@jnu.ac.in
* Correspondence: barbosa33@gmail.com; Tel.: +55-82-99999-3043

**Abstract:** Flash droughts (FDs) pose significant challenges for accurate detection due to their short duration. Conventional drought monitoring methods have difficultly capturing this rapidly intensifying phenomenon accurately. Machine learning models are increasingly useful for detecting droughts after training the models with data. Northeastern Brazil (NEB) has been a hot spot for FD events with significant ecological damage in recent years. This research introduces a novel 2D convolutional neural network (CNN) designed to identify spatial FDs in historical simulations based on multiple environmental factors and thresholds as inputs. Our model, trained with hydro-climatic data, provides a probabilistic drought detection map across northeastern Brazil (NEB) in 2012 as its output. Additionally, we examine future changes in FDs using the Coupled Model Intercomparison Project Phase 6 (CMIP6) driven by outputs from Shared Socioeconomic Pathways (SSPs) under the SSP5-8.5 scenario of 2024–2050. Our results demonstrate that the proposed spatial FD-detecting model based on 2D CNN architecture and the methodology for robust learning show promise for regional comprehensive FD monitoring. Finally, considerable spatial variability of FDs across NEB was observed during 2012 and 2024–2050, which was particularly evident in the São Francisco River Basin. This research significantly contributes to advancing our understanding of flash droughts, offering critical insights for informed water resource management and bolstering resilience against the impacts of flash droughts.

**Keywords:** flash drought; convolutional neural network; encoder–decoder architecture; Caatinga; climate change; hydro-climatic data

## 1. Introduction

Drought is a water shortage phenomenon caused by an imbalance in long-term water supply and demand [1,2]. According to a recent study, over 60% of the world's regions are affected by drought disasters annually, and the socio-economic and agricultural productions have been most affected [3]. In recent years, the academic community has shown that extremely high temperatures and rainfall deficits [4–6] can cause a very rapid onset and evolution of droughts, which have been referred to as flash droughts (FDs) [7–9]. Their identification and analysis are essential problems for risk management, informing governmental policy decisions, and advancing our fundamental understanding of the climate system. With human-induced climate change from increased $CO^2$ and other heat-trapping gases in the atmosphere, it is anticipated that the future will witness a progressive intensification and proliferation of climate extremes [8]. Among extreme events, FDs are ranked first in hazard characteristics given their severity on the ecological environment and their generated socio-economic losses [10].

The complex causes of FDs make their monitoring challenging. Although a common term, a universally accepted definition of what constitutes an FD does not exist (e.g., [11]). Nevertheless, from a broad physical standpoint, droughts are meso-climatic events that arise from interconnected atmospheric and hydrological processes [12], being self-supported by a positive feedback mechanism [13] (Figure 1). As soil moisture crosses a critical threshold, evapotranspiration potential rates decrease, lowering the atmospheric relative humidity and making the saturation point for rainfall harder to achieve. Overall, because of the stochastic nature of water demands across different regions of the world, a uniform operational definition of FD has become difficult to establish (e.g., [14]). Thus, rainfall is less likely, which exacerbates dryness.

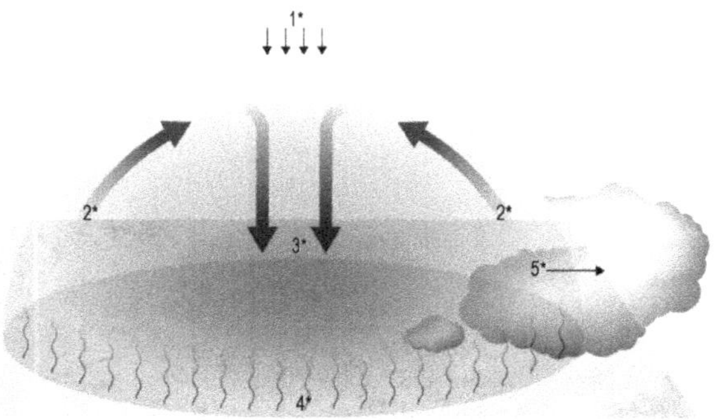


**Figure 1.** Schematic representation of a heat dome which involves high-pressure areas that trap and heat up the air below. This figure illustrates over a region that high pressure (**1 ***) in the atmosphere pushes warm air down toward the ground; hot air masses (**2 ***) expand vertically into the atmosphere as the air sinks (**3 ***) and it warms by compression; and the ground warms and loses its moisture, which makes it easier to heat even more (**4 ***). The dome of high pressure inhibits clouds and local rainfall and deflects away storms (**5 ***). These conditions lead to the onset of a flash drought, and consequences of the persistence of hydro-climatic anomalies over days or weeks can lead to substantial land degradation and desertification [4].

Commonly used drought indices in different types of gridded datasets are 3D climate model outputs from drought model simulations and 2D data from satellite retrievals. Drought indices have some limitations that make their performance region-specific [15]. Firstly, they are often defined with only one type of drought in mind. Secondly, they are local and do not account for spatiotemporal links. Thirdly, they assume specific probability distributions over the variables involved or simple thresholding ratios. Lastly, they are limited to inherent time scales. While they are simple to use, these limitations mean that they can lack the adaptability required to correctly identify the concept of FD, which shifts over time as climate change deviates climate conditions from normality. The complex causes of FDs make their detection and prediction systems challenging. Although considering convolutional neural networks (CNNs) can yield acceptable accuracy for classifying well-known types of extreme weather events, the choice of supervised learning can lead to variations in the effectiveness of data-driven approaches of drought events [16].

Nevertheless, the origins of Deep Neural Networks (DNNs) date back to the 1950s. The development and utilization of DNNs have recently accelerated rapid achievement of state-of-the-art results on classification, change and anomaly detection, forecasting, and model emulation tasks, among others [17–19]. For example, [20] recognizes their ability to learn optimal representations specifically from the data. Recent work has shown that fully supervised convolutional neural networks (CNNs) can yield acceptable accuracies for classifying well-known types of extreme weather events [21]. There is a specific benefit when applying the DNN algorithm to the detection of extremes. It helps to identify a specific type of extreme event from the anomalous data cloud and allows for better fitting of its distribution.

Deep learning is a new evolution of traditional machine learning research, designed to enable the computer to learn inherent characteristics of a dataset from a large pool of sample data, and classify and predict the newly received samples. CNNs are one of the more common deep learning architectures. They utilize layers of convolutional filters to automatically detect and learn hierarchical patterns and features from the input data. Through a combination of convolutional layers, pooling layers, and fully connected layers, CNNs effectively capture spatial and temporal dependencies. Thus, machine learning has been gradually applied to the study of drought detection because it can effectively deal with the nonlinear relationship between various drought factors [7,8]. However, given the novelty of the method and the challenges in implementation, there are few studies on drought monitoring using machine learning [20]. To overcome this shortcoming, this study develops a drought identification model using machine learning architecture, validates it with an independent set of data, and describes the spatial variability of flash drought events in northeastern Brazil (NEB). This model is based on a convolutional neural network (CNN) model. A comparative analysis with outcomes from the method's implementation relying on ground observations and satellite retrievals was carried out in 2012. In addition, the FD identification method was applied to the future emission scenario of CMIP6 (SSP5-8.5) to quantify the FD events occurring in NEB from 2024 to 2050. The research's contribution is introducing a convolutional encoder–decoder framework for identifying flash droughts.

The study is organized as follows. Section 2 describes the study area, datasets for the hydro-climatic variables, and drought index. Section 3 details the proposed spatial drought detection model. The evaluation of the results and discussions provided by the architecture in northeastern Brazil is presented in Sections 4 and 5, respectively. Finally, conclusions are drawn in Section 6.

## 2. Materials

### 2.1. Study Area

Northeastern Brazil (NEB) spans from approximately latitude 1.3° to 18.2° S and longitude 34.4° to 48.4° W, encompassing a land area of approximately 1.55 million square kilometers [4]. Located in northeastern South America, the region features both inland and coastal characteristics. The terrain slopes from high in the west to low in the east. The eastern area, being closer to the ocean, receives more precipitation and is relatively humid. The western region, situated inland and away from the ocean, is characterized by less precipitation and higher evapotranspiration, with more drought conditions compared to the east. Due to the complexity of the terrain and climate zones, along with a large population (53 million inhabitants), water scarcity is becoming increasingly severe, manifested in frequent drought events, significantly affecting agricultural development [22]. Furthermore, climate models described by Marengo et al. [23] suggest that the drought frequency in NEB will increase in the future due to a higher evaporative demand and persistent dry conditions associated with global warming. Consequently, flash droughts across NEB are also expected to increase in response to global warming, particularly as the region is an agriculture center where increasing $CO_2$ will impact plant growth [4], which is closely tied to the hydrologic cycle [24].

According to the Köppen–Geiger climate classification map published by Beck et al. [25], the study area mainly encompasses nine subtypes within three categories and its vegetation cover includes caatinga, rainforests, riparian forests, savannas, and montane forests, among others. Caatinga vegetation accounts for approximately 62% of NEB [26,27], with a rich diversity of vegetation types (Figure 2). The main plant species covered by Caatinga are Angico (*Anadenanthera colubrina*), Barriguda (*Ceiba glaziovii*), Cacto (*Cactaceae*), Carnaúba (*Copernicia prunifera*), Catingueira (*Caesalpinia pyramidalis*), Cumaru (*Amburana cearensis*), Facheiro (*Pilosocereus pachycladus*), and Juazeiro (*Ziziphus joazeiro*). These plant species withstand drought conditions and are used as food sources for humans and animals, and also play a crucial role in maintaining nutrient cycling within this biome [4]. Other climatic features of the study region are the annual mean rainfall of 400–800 mm, the annual average air temperature variation from 23 °C to 27 °C, and the potential evaporation (PET) of about 2000 mm year $^{-1}$ [26].

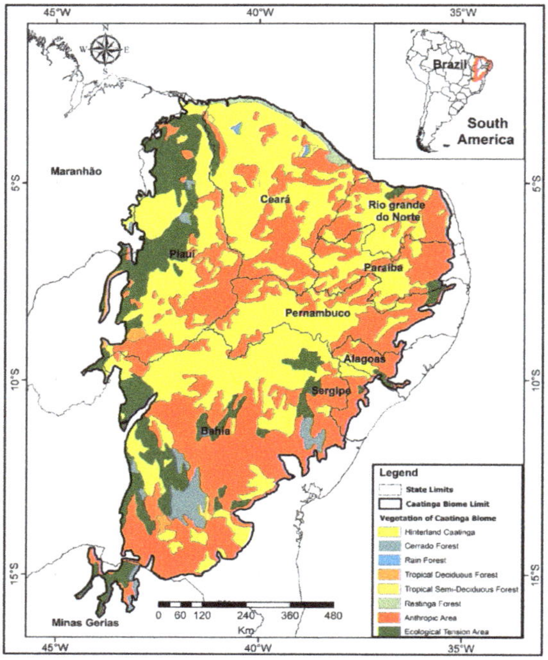


**Figure 2.** Caatinga vegetation map. It comprises the following states in Brazil: Alagoas (AL), Bahia (BA), Ceará (CE), Minas Gerais (MG), Maranhão (MA), Paraíba (PB), Piauí (PI), Pernambuco (PE), Rio Grande do Norte (RN), and Sergipe (SE).

*2.2. Data Sources*

We obtained hydro-climatic data from the Brazilian Daily Weather Gridded Data (BR-DWGD) developed by Xavier et al. [28], a data product derived from a collection of algorithms and computer software designed to interpolate and extrapolate from daily meteorological observations to produce gridded estimates of daily weather parameters. The BR-DWGD were used to assess and identify periods of flash drought across northeastern Brazil from 2010 to 2022. The precipitation (P) and potential evapotranspiration (PET) variables shared a common spatial grid of 0.1° with daily data. The BR-DWGD variables used for this assessment were (P) precipitation and potential evapotranspiration (PET), at a common spatial grid of 0.1° on a daily scale. Therefore, referring to the calculation method of the Standardized Precipitation Evapotranspiration Index (SPEI), this study calculated the water deficit based on P and PET data in northeastern Brazil, fitted the water deficit data series using the log-logistic method [29], and used python tools to calculate SPEI data on a 3-month scale. Daily surface soil moisture (SSM) data were obtained from the SMOS L3 SSM product provided by the Barcelona Expert Center over the period of 2010–2022. An overview of the retrieved values is provided by González-Zamora et al. [30]. For example, the correlation between P and SSM portrays a linearity (Figure 3), which guarantees a physical relationship between the two variables.

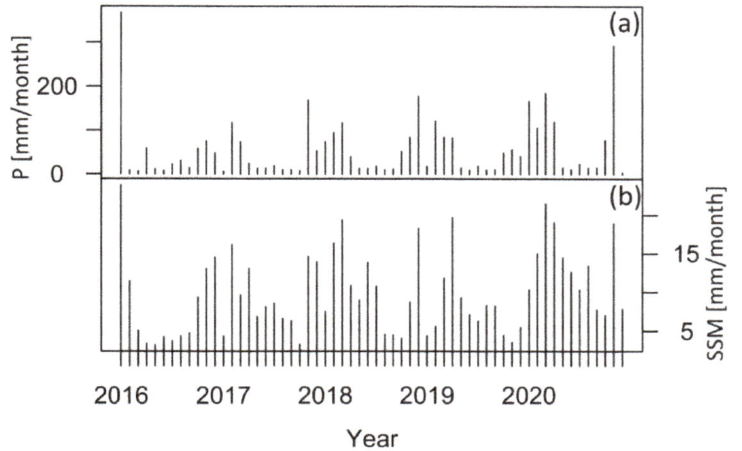


**Figure 3.** An example of (**a**) the observed precipitation (P) and (**b**) satellite retrievals for surface soil moisture (SMOS-based SSM data) obtained in an agricultural area across the study area. The linear correlation between these two variables is indicated by the $R^2 = 0.70$ from January 2016 to December 2020.

Daily NDVI data were provided by the Laboratory for Analyzing and Processing Satellite Images (LAPIS)'s archive [31], with a spatial resolution of 3 km. To synchronize the precipitation data's temporal scales (Figure 4), we generated monthly and annual NDVI data series. Subsequently, all raster datasets were resampled to a 5 km resolution to facilitate analysis and discussion, and details of each data source are provided in Table 1. Previous studies have shown that the SPEI data from the BR-DWGD have a strong relationship with the soil moisture and vegetation response in NEB [27,28].

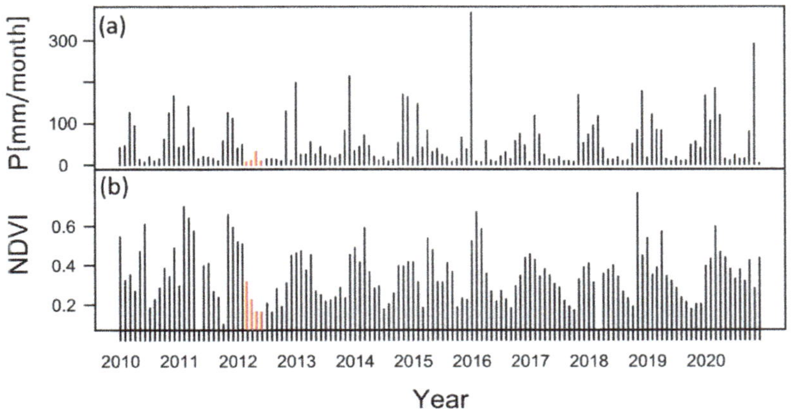


**Figure 4.** As in Figure 3 (**a**) the observed precipitation but referring to (**b**) NDVI [dimensionless]. The NDVI was derived from the MSG-SEVIRI product. The dry spell occurred in 2012 and is highlighted with red bars.

**Table 1.** Details of data sources.

| Product/Data Name | Time Period | Temporal Resolution | Spatial Resolution | Data Source | Accessed on |
|---|---|---|---|---|---|
| P | 2010 to 2020 | Daily | 0.1° | https://github.com/AlexandreCandidoXavier/BR-DWGD | 15 November 2023 |
| PET | 2010 to 2020 | Daily | 0.1° | | |
| SMOS L3 SSM (asc) | 2010 to 2020 | Daily | 0.225° | http://bec.icm.csic.es | 10 October 2023 |
| SMOS L3 SSM (des) | 2010 to 2020 | Daily | 0.225° | http://bec.icm.csic.es | 10 October 2023 |
| NDVI | 2010 to 2020 | Daily | 3 km | https://lapismet.com.br/ | 12 November 2023 |

For testing, in addition to the soil moisture validation dataset described in Table 1, we also ensured all data were at are at the same spatial resolution and scale. NDVI data were resampled to a 10 km resolution using the bilinear interpolation method in GEE (Google Earth Engine). The spatial resolution of the NDVI data was about 3 km.

## 3. Methods

### *3.1. The Standardized Precipitation Evapotranspiration Index (SPEI)*

In this study, the SPEI was computed as a daily-scale drought index, which considers both P and PET from the BR-DWGD. The SPEI computation involves fitting the cumulative moisture deficit (D) series with an appropriate probability distribution function, followed by normalization to derive SPEI categories [29]. A log-logistic distribution was then used to fit the D time series [32]. Additionally, the Kolmogorov–Smirnov (K-S) test with a 0.05 significance level was used to determine the optimal probability distribution for northeastern Brazil using the gma library in Python. Negative SPEI values indicate water deficit, while positive values denote surplus moisture (Figure 5). We employed the log-logistic probability distribution function to fit D and conducted SPEI calculations for northeastern Brazil using the gma library in Python.

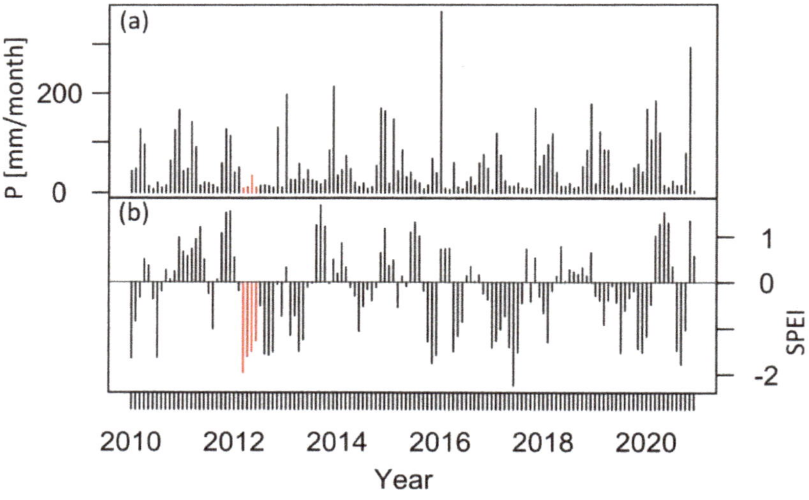


**Figure 5.** As in Figure 4 (**a**) the observed precipitation but referring to (**b**) the SPEI [dimensionless; base period January 2010–December 2020].

The PET is the sum of the amount of evaporation and transpiration from a reference vegetation of grass. It can be calculated with the Thornthwaite, Hargreaves, or Penman–Monteith equations [29]. The Thornthwaite equation is computed as follows:

$$\mathrm{PET} = 16\left(\frac{\mathrm{L}}{12}\right)\left(\frac{\mathrm{N}}{30}\right)\left(\frac{10\mathrm{Td}}{\mathrm{I}}\right)^{\alpha} \quad (1)$$

$$\alpha = \left(6.75 \times 10^{-7}\right)\mathrm{I}^3 - \left(7.71 \times 10^{-5}\right)\mathrm{I}^2 + \left(1.792 \times 10^{-2}\right)\mathrm{I} + 0.49239 \quad (2)$$

$$\mathrm{I} = \sum_{i=1}^{12}\left(\frac{\mathrm{Tmi}}{5}\right)^{1.514} \quad (3)$$

where PET is the estimated potential evapotranspiration (mm/month), Td denotes the mean monthly temperature (degrees Celsius), N is the number of days for each month, L indicates the local insolation (hours/month), and I is the annual heat index, which depends on the monthly mean temperatures (Tmi in degrees Celsius).

We analyzed the correlations between the monthly hydro-climatic variables' time series spanning from 2010 to 2022 and the SPEI sequence with various offsets (e.g., 0 months, 1 month, 2 months, and 3 months) over the study area. Utilizing an area-averaged time series of grid-scale R calculations, we detrended each variable to evaluate its response to drought (SPEI). Then, utilizing the maximum value, we identified the optimal response correlation for each variable and its corresponding lag intensity, employing a significance level of 95%. Subsequently, we applied a wavelet squared coherence analysis to reveal underlying oscillation patterns and changes in periodicities in a time–frequency domain between the area-averaged values of the SPEI against the SSM during their common time periods.

### 3.2. Model

The processing pipeline of the model is illustrated in Figure 6. First, given a set of hydro-climatic variables, we select spatial data for northeastern Brazil. During training, we select variable data around drought events (further details in Section 3.4). The resulting batch of selected variable defines the sample to feed the model. Then, a Spatial Convolutional Encoder–Decoder (S-CED) architecture processes this sample by extracting relevant SSM information (encoder) and transforming it into a probabilistic drought detection map through time (decoder). During training, the resulting map is compared with the ground truth map for drought detection to define the penalty by which the model is optimized. Later, the drought detection map through time is provided for the entire area of study by concatenating the sample in the sequential order it was taken. Adding convolutions to neural networks promotes two biases in the model: (i) nearby pixels (locations) are related to each other, and (ii) targets can appear anywhere in the input data [29].

The convolutional neural networks (CNNs) model can yield acceptable accuracy for classifying well-known types of extreme weather events. Further, it has been gradually applied to the study of drought construction because it can effectively deal with the nonlinear relationship between various drought factors [17]. This is consistent with what others have observed. Wei and Li [21] found that the overall performance of deep learning is comparable to that of a random forest, and its multi-layer method can find the best output in the case of high-dimensional data features. However, the pooling layer of the CNN model leads to the loss of some important features during training, which may be one of the reasons why the accuracy of the training model is lower than that of the random forest model.

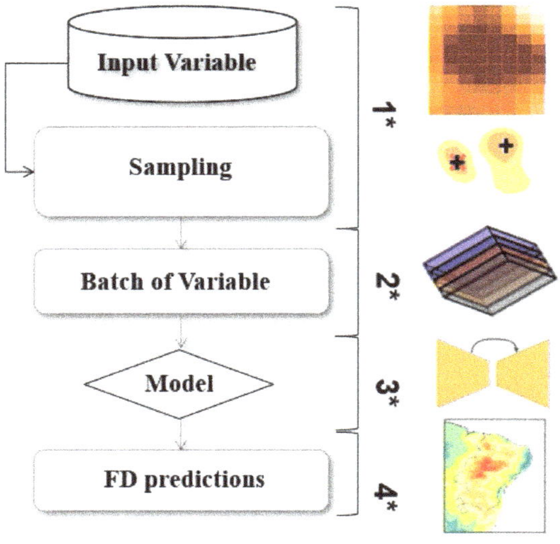


**Figure 6.** Schematic overview of the individual steps of the deep learning processing for FD detection. The input-selected hydro-climatic variable (**1** *) is taken by following a different sampling strategy for the training stages (**2** *). A batch of selected variables (**3** *) defines a sample received as input by the model, and a 2D convolutional encoder–decoder architecture provides probabilistic scores (**4** *) for an FD as its output.

### 3.3. Model Architecture Design

Detecting an FD requires network architectures capable of learning complex representations and exploiting interconnections in space and time. Adding convolutions to neural networks encourages two biases in the model: (i) nearby pixels (locations) are related to each other, and (ii) targets can appear anywhere in the input image [29]. Using convolutions significantly reduces the number of learnable parameters, resulting in faster learning and more scalable networks for deployment in real-case scenarios. Motivated by these properties, we propose an encoder–decoder architecture [33] that uses 2D convolutional layers to process hydro-climatic variables in space for FD detection as a binary classification task (FD vs. non-drought conditions) at the grid level. The splits are defined such that the amount of flash drought grids in each is similar, with the corresponding drought vs. non-drought ratios being 0.8%, excluding water. In a nutshell, the network receives as input a batch of data variables of size (lat, lon), with lat and lon being the latitude and longitude sizes, respectively. The model comprises two 2D convolutional layers for the encoder and two corresponding 2D convolutional layers for the decoder. CNNs utilize layers of convolutional filters to automatically detect and learn hierarchical patterns and features from the input data. Due to a reduction in the number of learnable parameters, deep learning offers high accuracy and efficiency in processing large datasets [34,35]. The architecture is shown in Figure 7.

The model uses a batch normalization, a normalization layer, a modified linear unit layer (ReLU), and a fully connected layer. At the encoder, max pooling is used after dropout layers to reduce the spatial dimensions by a factor of two. Likewise, residual skip connections [35,36] connect the encoder and the decoder blocks. At the top of the decoder, a final 2D convolutional layer followed by a sigmoid activation produces the class probabilities for each pixel, which constitute FD detection maps through time. In this study, we present the selected hydro-climatological data of size (240, 240) and output the probabilistic drought detection map of size (240, 240). Convolutional layers typically employ padding to extend the range of the convolution operation at image borders and produce an output that is the same size as the input [16,37]. Using valid convolutions has the noticeable caveat that each convolution produces feature maps with fewer elements. In a nutshell, the network receives as input a batch of selected data of variable size (lat, lon), with lat and lon being the latitude and longitude sizes, respectively. The model comprises 2D convolutional layers for the encoder and two corresponding 2D convolutional layers for the decoder. Exploiting convolutions significantly reduces the number of learnable parameters, resulting in faster learning and more scalable networks for deployment in real-case stages.

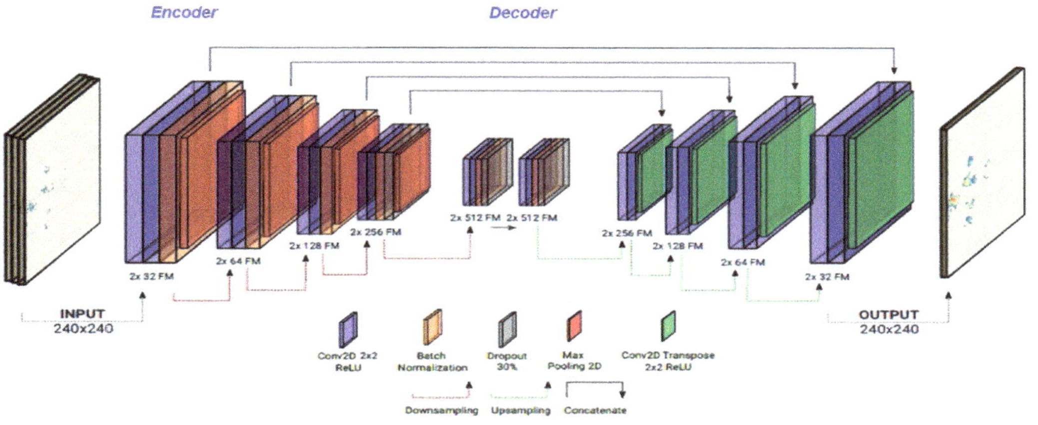


**Figure 7.** Schematic overview of the proposed SCED for drought detection. The model receives as input a selected data variable and provides a probabilistic FD detection map for multiple timesteps as its output. Represented by color are the 2D convolution + 2D batch normalization + LeakyReLU activation, 2D (spatial) pooling, 2D (spatial) upsampling, and 2D out convolution. At the bottom of each 2D convolution + 2D batch normalization layer + LeakyReLU activation, the output spatial size in terms of (lat, lon) is indicated. Skip connections are placed between the input of an encoder block and the activation layer of the corresponding decoder block.

### *3.4. Flash Drought Identification*

FDs were identified over the period 2010–2022 using SPEI and SSM values. The SPEI was spatially averaged for the study area. To enhance the accuracy of FD identification, the five-day average of the SSM time series was taken. The daily SPEI can exhibit noise due to short-term meteorological variability that can potentially obscure the FD onset [14,38]. By averaging the SPEI values over five-day intervals, the variability was mitigated while still being sufficient to capture the rapid intensification period characteristic of FDs. Flash drought events were identified using multi-criterion guidelines as defined in previous studies [4]. This method employs the following criteria: (1) the total decrease in the SPEI should be 2 or greater in a 30-day period; (2) at the end of the same 30-day period, the SPEI value should be less than $-1.5$ (or the SSM should be below the 20th percentile); and (3) after crossing the drought threshold value of $-0.5$, the SPEI should remain below $-0.5$ for at least 30 days. The first criterion captures the rapid intensification aspect of FDs and confirms that the identified events are not affected by temporary fluctuations in the SPEI or SSM due to increased rainfall, decreased temperatures, or increased cloud cover. The second and third criteria help separate FDs and dry spells and identify flash drought events that may have potential environmental impacts over more extended periods. The SSM is expressed in a percentage format. The daily values of the SSM data are assigned to different percentile categories of FDs, including 20th–25th, 15th–20th, 10th–15th, and <10th (Table 2). The quantile values provide a useful way of comparing the SPEI with varying orders of magnitude. They help to standardize the data and make them easier to compare and analyze FD categories across different regions. Although drought categories are binary, i.e., drought/non-drought, we rely on the assumption that the FD detection system should be continuous and capture the evolution of the phenomena. This is achieved by taking the probability distribution by which an example is labeled as belonging to one category and not to another (i.e., the similarity structure), as shown in Table 2.

**Table 2.** Thresholds used for flash drought classification.

| Drought Category | SPEI | Probability [%] [1] | SSM |
|---|---|---|---|
| Non-drought | >1.00 | >77.50 | >25th |
| Near normal (FD1) | 0.99 to −0.99 | 68.30 | 20th–25th |
| Moderate dry (FD2) | −1.00 to −1.49 | 9.20 | 15th–20th |
| Severe dry (FD3) | −1.50 to −1.99 | 4.40 | 10th–15th |
| Extreme dry (FD4) | <−2.00 | 2.30 | <10th |

[1] The cumulative probability of non-exceedance for each SPEI drought category.

Furthermore, the Coupled Model Intercomparison Project Phase 6 (CMIP6) was used to calculate the SPEI3 for identifying future stages of FD based on the SPEI under a high-emission scenario. The following variables were employed: water evapotranspiration flux, maximum air temperature, minimum air temperature, and precipitation. These data were derived from historical simulations and Shared Socioeconomic Pathways (SSPs) that characterize four future scenarios; we chose to use the SSP5-8.5 of CMIP6. The historical simulation data cover the period from 2010 to 2015, while the data for the future scenario cover the period from 2024 to 2050. We used the NorESM Climate Modeling Consortium (NCC) agency, which has a spatial resolution of 1.25°.

## 4. Results

### *4.1. Evaluation of Hydro-Climatic Data in Response to Drought*

In the analysis of wavelet coherence between the SPEI and SSM (Figure 8a), it is evident that the periodicity of 8–12 months is predominantly high from 2010 to 2016 (significant at a 95% level). The cone area denotes the cone of influence under which the period (month) can be considered for the analysis. Hence, it is evident from the wavelet analysis that the SPEI changes are in accord with the different periodicities of the SSM, which are also interpreted as drought-induced temporal effects on the SSM over NEB. Considering that the SPEI has a stronger explanatory power on the intensity of drought, which includes information on temperature, precipitation, and potential evapotranspiration, this study selected the SSM, NDVI, and PET as independent hydro-climate variables.

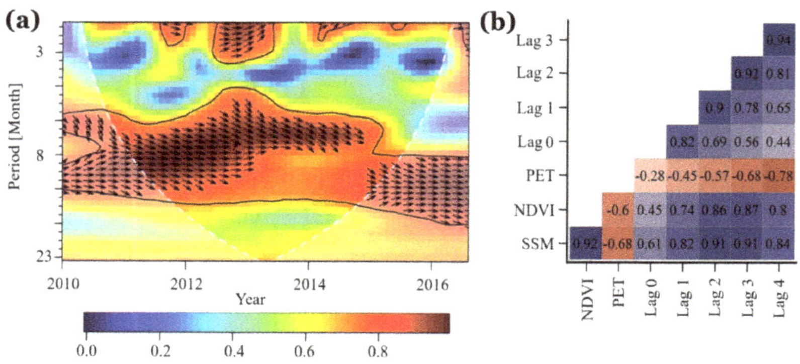


**Figure 8.** Results of the analysis on the driving hydro-climatic data of temporal responses over the entirety of NEB during 2010–2016. (**a**) The squared wavelet coherence of the SPEI against the SSM, and (**b**) the lagged correlations between the SPEI and the variables involved (SSM, NDVI, and PET).

Figure 8b reflects the interactive effects of each variable on the temporal impact of drought (SPEI). The SSM and NDVI dominate as the driving factors, followed by the PET. We identified the lag effect durations of each variable during drought as follows: SSM > NDVI > PET. In addition, we observed significant negative correlations between the PET and drought, indicating a stronger response of the PET to lagging drought (i.e., higher sensitivity). Longer lag effect times imply that the SSM is more likely to be affected by the lingering effects of previous months' droughts and is less likely to recover from earlier droughts. Compared to other commonly used meteorological drought indices, the SPEI exhibits a greater capability to measure the impacts of drought on agricultural and ecological responses [27,29]. Thus, we conclude that the SSM was the optimal variable for drought detection.

### *4.2. Identifying and Mapping Flash Drought Events*

To showcase the ability of the SSM to capture FDs, the well-studied 2012 NEB event was examined. This drought event occurred during the autumn, winter, and spring of 2012 and was a historically unprecedented event in NEB, ranked as the third-largest drought in terms of aerial extent since 1901. The event was preceded by La Niña conditions during the summers of 2011–2012, resulting in drier-than-normal conditions at the beginning of the year [26,31]. Figure 9 shows the rapid onset and progression of the 2012 flash drought demonstrated by SSM values every 5 days (i.e., pentads) from 1 March to 28 December. The drought reached its peak severity at the end of August, with nearly three-fourths of NEB classified as having severe or extreme drought conditions according to the SPEI categories. The mean and median lengths of the flash drought events in 2012 were 24 and 114 days. This indicates that the flash drought events identified in NEB lasted approximately 1–4 months.

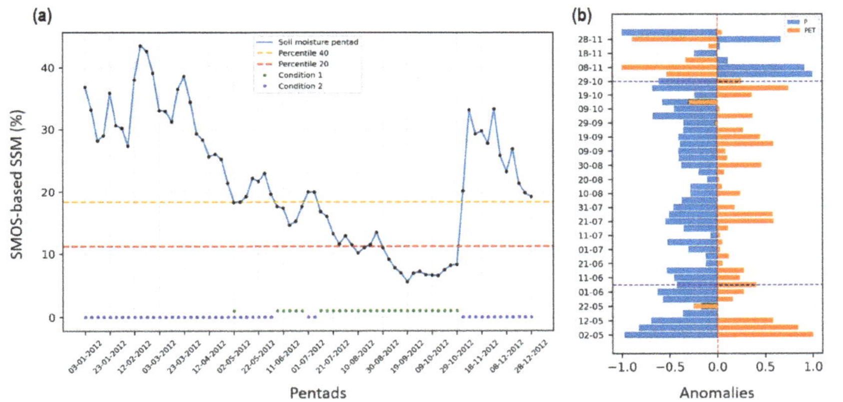


**Figure 9.** (**a**) Time series of the five-day moving average of the daily SMOS-based SSM values over the entirety of NEB from 1 March to 28 December 2012. The dashed orange line and dashed red line represent the 40th and 20th percentiles for soil moisture, respectively. The blue dot represents non-FD (condition 1), and the green dot represents FD (condition 2). (**b**) The blue–orange vertical time series of anomaly values represents the period of precipitation (P) and potential evapotranspiration (PET) data from 2 May to 28 November 2012. The vertical dashed blue lines indicate the period between the FD conditions.

By examining Figure 9, we see that the minimum SSM values were experienced in August–October, reaching less than the 10th percentile. The synchronization of the P and PET during the onset of the 2012 flash drought highlights the connection between the onset of flash drought conditions indicated by the SSM and subsequent reductions in precipitation, offering insights into the linked dynamics of the atmospheric and hydrological responses.

The analysis of FD events in 2012 within NEB reveals variations in the visual comparative analysis between SMOS-based SSM retrievals and the CNN model simulation. The employed criteria ensured robust identification of FDs by considering the multi-criterion approach outlined in Section 3.4. Figure 10 shows the five-day averages of the SSM and CNN from 1 March to 28 December 2012. Each map is represented by a mean grid-point (pixel level), with red indicating instances where the SSM caused an FD, and green indicating non-drought locations. Non-drought conditions fall within typical hydro-climatic conditions and are comparable to easy-to-classify locations. Moreover, FDs have various stages in their development, as shown in Figure 9a.

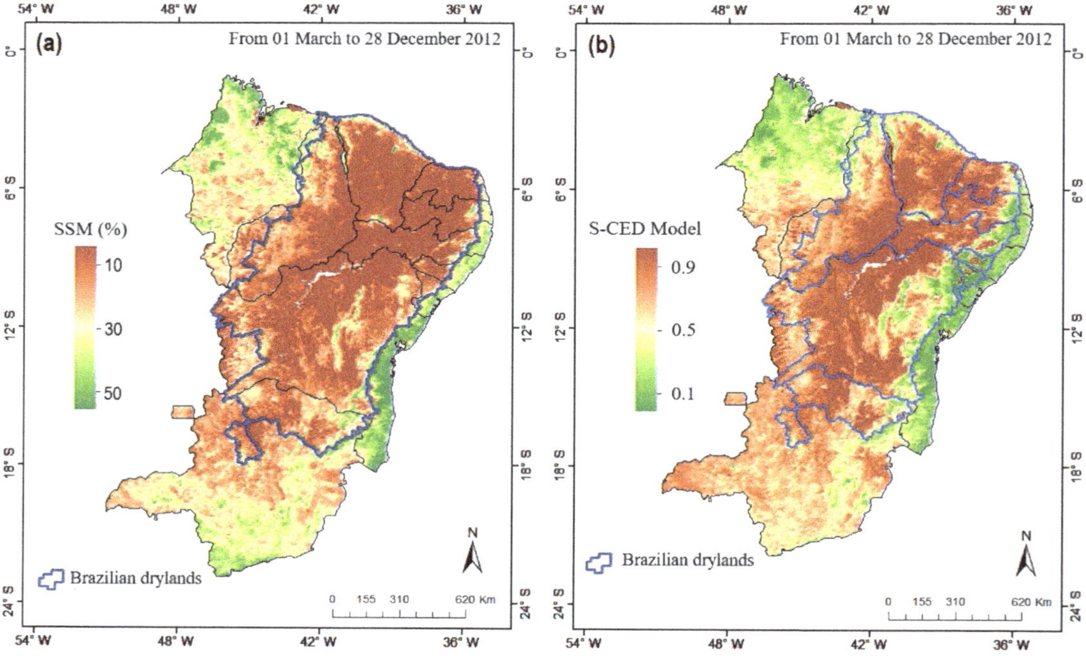


**Figure 10.** Visual comparison of the performance of (**a**) the SMOS-based SMOS L3 SSM product and (**b**) the S-CED model trained with SSM data for the 2012 FD events identified from 1 March to 28 December 2012. The red color represents the higher scores of the model trained with the SSM. In contrast, green means lower values of the model. The Brazilian drylands boundary is displayed as the blue line (left side), and northeastern Brazil and its states in blue lines (right side).

We provide a visual comparison of the performance of the S-CED model trained with the SSM and SMOS L3 SSM product. First, in Figure 10, we present the northeastern Brazil probabilistic drought detection map provided by the S-CED model during the 2012 FD events. A more intense red color in the S-CED model corresponds to higher scores. This visualization is used to identify the model's distribution results of all the grids in the study region compared to soil moisture conditions estimated from the SSM product. The red areas in the maps delimit the areas of flash drought events identified for the region, and a green colormap is used to represent the non-drought category. Thus, the probabilistic scores of the S-CED model in non-drought category areas (the green color) are higher than those obtained by the SSM product. The blue contours encompass the Brazilian drylands, where the visual accuracy is higher between two maps. We also obtain clear signals of potential non-droughts, highlighting the usefulness of the proposed model and the benefits of the method for flash drought identification.

FD events were identified in the growing season (February–May) from 2024 to 2050 within NEB using the CMIP6 predictions. Our focus was on capturing the FD events to better comprehend their impact on the main rainy season of NEB. Figure 11 shows the intensity (percentile/year) of FDs in the SSP5-8.5 scenario. The primary impact is primarily concentrated in the São Francisco River Basin, within NEB's semi-arid region. The entire basin is engulfed in FD conditions, characterized by increased severity in its central region, where the dominant FD4 and FD3 are observed. These categories indicate the severity of FD conditions, which are close to the extreme drought category defined by the SPEI, and pose significant challenges to water resources, agriculture, and ecosystems within those areas. Additionally, the results revealed the FD intensity, showing that 28% of the region was categorized as FD1. FD2 accounted for 23%, while FD3 and FD4 comprised 21% of the NEB-FD area. It indicates that NEB's semi-arid region (i.e., the Caatinga biome) will face significant challenges in terms of FD risk under future scenarios. Therefore, understanding the interconnected nature of water management within the entire basin is crucial for comprehensively assessing and addressing the impacts of severe flash droughts in those areas.

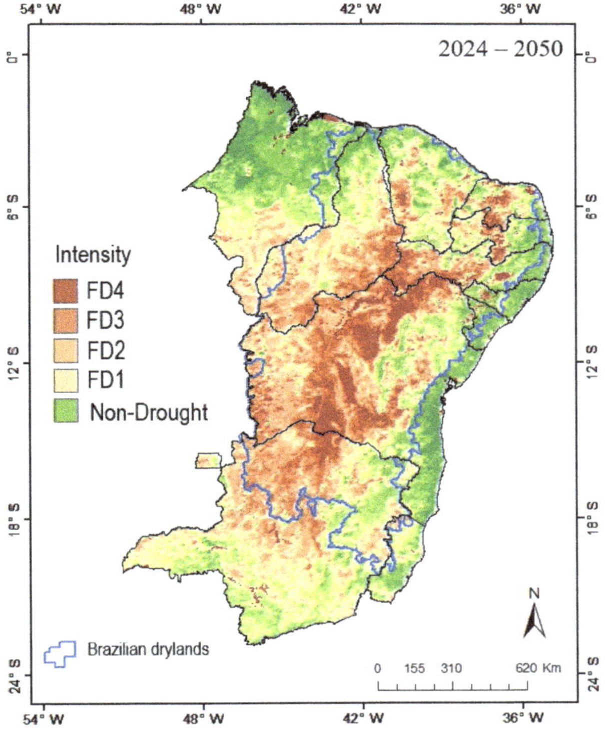


**Figure 11.** FD severity in SSP-8.5 identified in the growing season (February–May) from 2024 to 2050 within northeastern Brazil. Brazilian drylands boundary denoted by blue line.

## 5. Discussion

We present a new method for FD identification using different types of input data. An FD is a rapid drought in a short period caused by severe heat waves and rainfall deficiency [4,25]. It is very difficult to accurately detect an FD [7,12]. The occurrence of an FD is related to the complex interaction of soil moisture, evapotranspiration, and vegetation. This research adopts the CNN architecture for FD identification to output data from model simulations in northeastern Brazil (NEB) by integrating multiple sources of hydrometeorological data that represent the atmosphere (P and PET data), surface (NDVI data), and subsurface (SSM data). We selected SSM data to train the proposed model using a cross-correlation analysis and tried to avoid splitting the responses of multiple drought types. The complete area was divided into data of the same size as the samples used for training, and the model generated probabilistic scores for surface soil moisture data. These scores were then used to reconstruct the entire region.

The method-based neural representation has been applied in northeastern Brazil, and it can identify an FD after training with SSM data. The comparison against outcomes from the CNN model implementation reveals a slight increase in its capability to properly identify FD areas, which is more substantial when forcing the algorithm with surface soil moisture. Examination of the CNN's ability to effectively capture FD events revealed its proficiency in identifying the development phase. The spatial mapping of the 2012 FD showcased variations in intensity across NEB, with distinct patterns in different regions. The southwest NEB region experiences frequent FDs, while the northeast encounters less frequent events, and its semi-arid region faces severe FD conditions. Additionally, our study underscores the lag effects of drought on different environmental factors, enriching our comprehension of soil moisture's response to drought. The delayed response emphasized the complex relationship between meteorological and hydrological factors during FD events, highlighting the challenge of predicting and mitigating the consequences of FDs on the NEB drylands.

The NEB drought events of 2012 showed an evident accuracy in the spatial domain, as shown in Figure 7. The soil moisture dataset from satellite observations is at the root of it. We can see how the learned models obtain distinctive simulations for the registered FD events. We also find clear signals of potential unregistered droughts, highlighting the usefulness of the proposed model and the benefits of the method for drought detection. CNNs offer high accuracy and efficacy in processing large datasets, though they may lose some features due to pooling layers, slightly lowering the accuracy compared to the random forest model [39].

We found that the model had enough representational capacity with two layers for the encoder and two for the decoder. More layers decreased the performance results for the validation data. Convolutional layers typically employ padding to extend the range of the convolution operation at image borders and produce an output that is the same size as the input [39,40]. Multiple values for the padding exist, with the zero-valued one being the most common. However, padding introduces artificial distortions that do not align with the actual behavior of hydro-climatological variables. As such, we refrained from using padding in the convolutions and performed valid convolutions. Using valid convolutions has the noticeable caveat that each convolution produces feature maps with fewer elements.

Nevertheless, uncertainties persist. Comparing the CNN with the surface soil moisture provided a holistic perspective on the interconnected dynamics during FD events. The synchronization observed between the severity of FD conditions indicated by the CNN and subsequent reductions in soil moisture highlighted the potential of our proposed approaches to improve FD detection algorithms. The uncertainty and bias in the SSM data are at the root of unregistered FD events. We acknowledge that validating the model under categories of drought and non-drought settings is challenging since these problems also affect the validation data. This can lead to poor estimation of generalization capabilities and wrongly tuned models [41]. Since the number of drought and non-drought grids varies considerably, we balanced their contribution by computing a correction factor for each location based on [41].

By leveraging the CMIP6 (SSP5-8.5) as a predictive tool, our findings revealed that the FD severity in the São Francisco River Basin within NEB will increase in the future due to higher persistent dry conditions associated with global warming. In general, an expansion of the area under FD events in the context of high-emission scenarios was observed the middle and south areas of the basin from 2024 to 2050 (Figure 11). By examining Figure 11, we see that a wide range of minimum SPEI values are observed across NEB, spanning from 0.99 to <−2.00 (see Table 1). These values indicated the severity of flash drought conditions, with lower SPEI values representing more intense flash droughts (FD4). NEB's areas with minimum SPEI values higher than −2.00 are likely to experience more extreme flash drought events, posing significant challenges to water resources, agriculture, and ecosystems within those areas.

Nevertheless, as shown in Figure 11, the high dependence of the NEB rainy season (February–May) on weather conditions with insufficient rainfall has raised concern about the risk climate change poses to the entire regional economy [26]. These concerns are even greater since climate change projections suggest that future flash droughts will be more severe. In the context of the growing season, agricultural activities that are dependent on consistent water availability face disruptions that can impact crop yields and overall food production [23]. Changes in plant growth and transpiration due to climate change will have a direct impact on watershed processes, potentially leading to an increased intensity of drought conditions in the São Francisco River Basin in northeastern Brazil [26]. However, because of the uncertainties in the independent climate models of CMIP6, multi-model ensembles were typically used to minimize the impact of the models' uncertainties in the prediction results.

## 6. Conclusions

Flash drought (FD) detection and its prediction is of utmost importance in the present scenario because of its increase due to climate change. Our research introduces a novel methodology based on a deep learning model which, after training, can identify an FD. This process embeds spatiotemporal FD-identifying thresholds into the process of generating probabilistic scores from models that are trained with surface soil moisture data. In conclusion, the goals of the study were met by introducing a 2D convolutional encoder–decoder architecture and integrating multiple sources of hydro-climatic data to identify the complexity of spatial FD events across northeastern Brazil. As flash droughts pose challenges to accurate detection due to their complex spatial–temporal features, the insights gained from this research provide a foundation for further studies.

**Author Contributions:** H.A.B., methodology, formal analysis, investigation, writing—original draft, visualization, conceptualization, funding acquisition, and project administration. C.O.B., conceptualization, writing—review, and editing. T.V.L.K., writing—review and editing. All authors have read and agreed to the published version of the manuscript.

**Funding:** This study was supported by the Coordenação de Aperfeiçoamento de Pessoal de Nível Superior (CAPES 01/2022), Brazil, under the Project No. 88881.7050501/2022-01 to H.A.B through PEPEEC (Programa Emergencial de Prevenção e Enfretamento de Desastres Relacionados a Emergências Climáticas, Eventos Extremos e Acidentes Ambientais). It also had the support of the CNPq under project No. 403223/2021-0 to H.A.B through the Desertification Monitoring Program in the Brazilian Semi-arid Region.

**Institutional Review Board Statement:** Not applicable.

**Informed Consent Statement:** Not applicable.

**Data Availability Statement:** Publicly available datasets were analyzed in this study. This data can be found here: https://www.lapismet.com.br.

**Acknowledgments:** The authors are grateful for the comments from three reviewers who helped to significantly improve the quality of the manuscript. The authors offer thanks for the use of the European Weather Cloud (EWC) for providing processing capabilities for large data.

**Conflicts of Interest:** The authors declare that they have no known competing financial interests or personal relationships that could have appeared to influence the work reported in this paper.

## References


1. Wilhite, D.A. *Chapter I Drought as a Natural Hazard: Concepts and Definitions*; Drought A Glob. Assess; Drought Mitigation Center Faculty Publications: Lincoln, NE, USA, 2000.
2. Cook, B.I.; Mankin, J.S.; Anchukaitis, K.J. Climate Change and Drought: From Past to Future. *Curr. Clim. Chang. Rep.* **2018**, *4*, 164–179. [CrossRef]
3. Wu, C.; Ning, S.; Jin, J.; Zhou, Y.; Zhou, L.; Bai, X.; Zhang, L.; Cui, Y. Construction and application of comprehensive drought index based on uncertainty cloud reasoning algorithm. *Sci. Total Environ.* **2021**, *779*, 146533. [CrossRef]
4. Barbosa, H.A. Understanding the rapid increase in drought stress and its connections with climate desertification since the early 1990s over the Brazilian semi-arid region. *Arid Environ.* **2024**, *222*, 105142. [CrossRef]
5. Ahmad, S.K.; Kumar, S.V.; Lahmers, T.M.; Wang, S.; Liu, P.-W.; Wrzesien, M.L.; Bindlish, R.; Getirana, A.; Locke, K.A.; Holmes, T.R.; et al. Flash Drought Onset and Development Mechanisms Captured with Soil Moisture and Vegetation Data Assimilation. *Water Resour. Res.* **2022**, *58*, e2022WR032894. [CrossRef]
6. Lisonbee, J.; Woloszyn, M.; Skumanich, M. Making sense of flash drought: Definitions, indicators, and where we go from here. *J. Appl. Serv. Climatol.* **2021**, *2021*, 1–19. [CrossRef]
7. Svoboda, M.; LeComte, D.; Hayes, M.; Heim, R.; Gleason, K.; Angel, J.; Rippey, B.; Tinker, R.; Palecki, M.; Stooksbury, D.; et al. THE DROUGHT MONITOR. *Bull. Am. Meteorol. Soc.* **2022**, *83*, 1181–1190. [CrossRef]
8. Yuan, X.; Ma, F.; Li, H.; Chen, S. A review on multi-scale drought processes and prediction under global change. *Trans. Atmos. Sci.* **2020**, *43*, 225–237.
9. Mukherjee, S.; Mishra, A.K. A Multivariate Flash Drought Indicator for Identifying Global Hotspots and Associated Climate Controls. *Geophys. Res. Lett.* **2022**, *49*, 2. [CrossRef]
10. Ji, Y.; Fu, J.; Lu, Y.; Liu, B. Three-dimensional-based global drought projection under global warming tendency. *Atmos. Res.* **2023**, *291*, 106812. [CrossRef]
11. Xu, Y.; Zhang, X.; Hao, Z.; Hao, F.; Li, C. Projections of future meteorological droughts in China under CMIP6 from a three-dimensional perspective. *Agric. Water Manag.* **2021**, *252*, 106849. [CrossRef]
12. Zargar, A.; Sadiq, R.; Naser, B.; Khan, F.I. A review of drought indices. *Environ. Rev.* **2011**, *19*, 333–349. [CrossRef]
13. Zhou, S.; Williams, A.P.; Berg, A.M.; Cook, B.I.; Zhang, Y.; Hagemann, S.; Lorenz, R.; Seneviratne, S.I.; Gentine, P. Land–atmosphere feedbacks exacerbate concurrent soil drought and atmospheric aridity. *Proc. Natl. Acad. Sci. USA* **2019**, *116*, 18848–18853. [CrossRef]
14. Zhang, Q.; Shi, R.; Singh, V.P.; Xu, C.Y.; Yu, H.; Fan, K.; Wu, Z. Droughts across China: Drought factors, prediction and impacts. *Sci. Total Environ.* **2022**, *803*, 150018. [CrossRef]
15. Quiring, S.M.; Ford, T.W.; Wang, J.K.; Khong, A.; Harris, E.; Lindgren, T.; Goldberg, D.W.; Li, Z. The north American soil moisture database development and applications. *Bull. Am. Meteorol. Soc.* **2016**, *97*, 1441–1460. [CrossRef]
16. Chattopadhyay, A.; Hassanzadeh, P.; Pasha, S. Predicting clustered weather patterns: A test case for applications of convolutional neural networks to spatio-temporal climate data. *Sci. Rep.* **2020**, *10*, 1317. [CrossRef]
17. Reichstein, M.; Camps-Valls, G.; Stevens, B.; Jung, M.; Denzler, J.; Carvalhais, N.; Prabhat, F. Deep learning and process understanding for data-driven earth system science. *Nature* **2019**, *566*, 195–204. [CrossRef]
18. Camps-Valls, G.; Tuia, D.; Zhu, X.X.; Reichstein, M. *Deep Learning for the Earth Sciences: A Comprehensive Approach to Remote Sensing, Climate Science and Geosciences*; John Wiley & Sons: Hoboken, NJ, USA, 2021.
19. Vandal, T.; Kodra, E.; Ganguly, S.; Michaelis, A.; Nemani, R.; Ganguly, A.R. Generating high resolution climate change projections through single image super-resolution: An abridged version. *IJCAI Int. Jt. Conf. Artif. Intell.* **2018**, 5389–5393. [CrossRef]
20. Gibson, P.B.; Chapman, W.E.; Altinok, A.; Monache, L.D.; DeFlorio, M.J.; Waliser, D.E. Training machine learning models on climate model output yields skillful interpretable seasonal precipitation forecasts. *Commun. Earth Environ.* **2021**, *2*, 159. [CrossRef]
21. Wei, T.; Li, Y.F. Does tail label help for large-scale multi-label learning? *IEEE Trans. Neural Netw. Learn. Syst.* **2020**, *31*, 2315–2324.
22. IBGE Instituto Brasileiro de Geografia e Estatística. População. Estados. Rio de Janeiro. *IBGE* **2023**. Available online: http://censo2023.ibge.gov.br/apps/atlas/ (accessed on 20 April 2024).
23. Marengo, J.A.; Torres, R.R.; Alves, L.M. Drought in Northeast Brazil-past, present, and future. *Theor. Appl. Climatol.* **2017**, *129*, 1189–1200. [CrossRef]
24. Barbosa, H.A. Flash drought and its characteristics in northeastern South America during 2004–2022 using satellite-based products. *Atmosphere* **2023**, *14*, 1629. [CrossRef]
25. Beck, H.E.; McVicar, T.R.; Vergopolan, N.; Berg, A.; Lutsko, N.J.; Dufour, A.; Zeng, Z.; Jiang, X.; van Dijk, A.I.J.M.; Miralles, D.G. High-resolution (1 km) Köppen-Geiger maps for 1901–2099 based on constrained CMIP6 projections. *Sci. Data* **2023**, *10*, 724. [CrossRef]
26. Buriti, C.; Humberto, A.B.; Franklin, P.; Kumar, T.V.L.; Manoj, K.T.; Koteswara, R. Un siglo de sequías: ¿por qué las políticas de agua no desarrollaron la región semiárida brasileña? *Rev. Bras. De Meteorol.* **2020**, *35*, 683–688. [CrossRef]
27. Giovannettone, J.; Franklin, P.; Humberto, B.A.; Carlos, A.C.; Kumar, T.V.L. Characterization of links between hydro-climate indices and long-term precipitation in Brazil using correlation analysis. *Int. J. Climatol.* **2020**, *40*, 5527–5541. [CrossRef]
28. Xavier, A.C.; Scanlon, B.R.; King, C.W.; Alves, A.I. New improved Brazilian daily weather gridded data (1961–2020). *Int. J. Climatol.* **2022**, *42*, 8390–8404. [CrossRef]

29. Vicente-Serrano, S.M.; Beguería, S.; Lopez-Moreno, J.I. A multiscalar drought index sensitive to global warming: The Standardized Precipitation Evapotranspiration Index. *J. Clim.* **2010**, *23*, 1696–1718. [CrossRef]
30. González-Zamora, Á.; Sánchez, N.; Martínez-Fernández, J.; Gumuzzio, Á.; Piles, M.; Olmedo, E. Long-term SMOS soil moisture products: A comprehensive evaluation across scales and methods in the Duero Basin (Spain). *Phys. Chem. Earth Parts A/B/C* **2015**, *83–84*, 123–136. [CrossRef]
31. Barbosa, H.A.; Lakshmi Kumar, T.V. Influence of rainfall variability on the vegetation dynamics over Northeastern Brazil. *J. Arid Environ.* **2016**, *124*, 377–387. [CrossRef]
32. Stagge, J.H.; Tallaksen, L.M.; Gudmundsson, L.; Van Loon, A.F.; Stahl, K. Candidate distributions for climatological drought indices (SPI and SPEI). *Int. J. Climatol.* **2015**, *35*, 4027–4040. [CrossRef]
33. Sun, T.; Ferreira, V.; He, X.; Andam-Akorful, S. Water Availability of São Francisco River Basin Based on a Space-Borne Geodetic Sensor. *Water* **2016**, *8*, 213. [CrossRef]
34. Shen, R.; Huang, A.; Li, B.; Guo, J. Construction of a drought monitoring model using deep learning based on multi-source remote sensing data. *Int. J. Appl. Earth Obs. Geoinf.* **2019**, *79*, 48–57. [CrossRef]
35. Ji, S.; Xu, W.; Yang, M.; Yu, K. 3d convolutional neural networks for human action recognition. *IEEE Trans. Pattern Anal. Mach. Intell.* **2012**, *35*, 221–231. [CrossRef]
36. He, K.; Zhang, X.; Ren, S.; Sun, J. Deep residual learning for image recognition. In Proceedings of the 2016 IEEE Conference on Computer Vision and Pattern Recognition (CVPR), Las Vegas, NV, USA, 27–30 June 2016; pp. 770–778.
37. Khosravi, K.; Panahi, M.; Golkarian, A.; Keesstra, S.D.; Saco, P.M.; Bui, D.T.; Lee, S. Convolutional neural network approach for spatial prediction of flood hazard at national scale of iran. *J. Hydrol.* **2020**, *591*, 125552. [CrossRef]
38. He, M.; Kimball, J.S.; Yi, Y.; Running, S.; Guan, K.; Jensco, K.; Maxwell, B.; Maneta, M. Impacts of the 2017 flash drought in the US Northern plains informed by satellite-based evapotranspiration and solar-induced fluorescence. *Environ. Res. Lett.* **2019**, *14*, 074019. [CrossRef]
39. Prodhan, F.A.; Zhang, J.; Hasan, S.S.; Pangali Sharma, T.P.; Mohana, H.P. A review of machine learning methods for drought hazard monitoring and forecasting: Current research trends, challenges, and future research directions. *Environ. Model. Softw.* **2022**, *149*, 105327. [CrossRef]
40. Zhang, A.; Lipton, Z.C.; Li, M.; Smola, A.J. *Dive into Deep Learning*; Cambridge University Press: Cambridge, UK, 2023.
41. Lawrence, S.; Burns, I.; Back, A.; Tsoi, A.C.; Giles, C.L. Neural network classification and prior class probabilities. In *Neural Networks: Tricks of the Trade*; Springer: Berlin/Heidelberg, Germany, 2012; pp. 295–309.

*atmosphere*

MDPI

*Article*

# Basin-Scale Daily Drought Prediction Using Convolutional Neural Networks in Fenhe River Basin, China

**Zixuan Chen [1], Guojie Wang [2,*], Xikun Wei [1,3], Yi Liu [4], Zheng Duan [5], Yifan Hu [1] and Huiyan Jiang [1]**

1 Collaborative Innovation Center on Forecast and Evaluation of Meteorological Disasters, Nanjing University of Information Science and Technology, Nanjing 210044, China; chen_zixuan@nuist.edu.cn (Z.C.); xikunw@nuist.edu.cn (X.W.); huyifan@nuist.edu.cn (Y.H.); huiyanjiang@nuist.edu.cn (H.J.)

2 School of Remote Sensing and Geomatics Engineering, Nanjing University of Information Science and Technology, Nanjing 210044, China

3 Department of Engineering Hydrology and Water Management, Technical University of Darmstadt, 64287 Darmstadt, Germany

4 School of Civil and Environmental Engineering, University of New South Wales, Sydney 2052, Australia; yi.liu@unsw.edu.au

5 Department of Physical Geography and Ecosystem Science, Lund University, SE-22362 Lund, Sweden; zheng.duan@nateko.lu.se

* Correspondence: gwang@nuist.edu.cn

**Abstract:** Drought is a natural disaster that occurs globally and can damage the environment, disrupt agricultural production and cause large economic losses. The accurate prediction of drought can effectively reduce the impacts of droughts. Deep learning methods have shown promise in drought prediction, with convolutional neural networks (CNNs) being particularly effective in handling spatial information. In this study, we employed a deep learning approach to predict drought in the Fenhe River (FHR) basin, taking into account the meteorological conditions of surrounding regions. We used the daily SAPEI (Standardized Antecedent Precipitation Evapotranspiration Index) as the drought evaluation index. Our results demonstrate the effectiveness of the CNN model in predicting drought events 1~10 days in advance. We evaluated the predictions made by the model; the average Nash–Sutcliffe efficiency (NSE) between the predicted and true values for the next 10 days was 0.71. While the prediction accuracy slightly decreased with longer prediction lengths, the model remained stable and effective in predicting heavy drought events that are typically difficult to predict. Additionally, key meteorological variables for drought predictions were identified, and we found that training the CNN model with these key variables led to higher prediction accuracy than training it with all variables. This study approves an effective deep learning approach for daily drought prediction, particularly when considering the meteorological conditions of surrounding regions.

**Keywords:** drought; prediction; deep learning; CNN

## 1. Introduction

Drought, which has been exacerbated by climate change, is currently one of the most complex natural disasters with a significant global impact [1–4]. Climate change has led to rising global temperatures and decreasing precipitation, resulting in drought as a major natural disaster worldwide. Prolonged droughts can make soil moisture deficient, which in turn seriously threatens food security [5]. China has been significantly impacted by drought, with increasing losses threatening agricultural production and socio-economic development [6–9]. The Huang–Huai–Hai River (HHH) basin, in particular, is vulnerable to drought due to the combined effects of climate change and human activities [10,11].

The early warning and accurate assessment of drought is crucial for the effective mitigation of its damage [12,13]. Various methods have been developed for drought prediction, including traditional and machine learning-based approaches. Early approaches include Autoregressive Integrated Moving Average (ARIMA) and Multiplicative Seasonal

Autoregressive Integrated Moving Average (SARIMA), which are good at dealing with more complex time series problems and can take seasonal factors into account when making drought predictions [14]. Li et al. [15] developed a physical–empirical prediction model for predicting drought in northeastern China. However, due to the limitation of computer performance, the prediction accuracy obtained by previous methods is generally low.

In recent years, there has been a surge in the development of machine learning-based drought prediction models, resulting in a significant improvement in their predictive accuracy [16]. Feng et al. [17] used three machine learning methods, namely bias-corrected random forest (BRF), support vector machine (SVM), and the multi-layer perceptron neural network (MLP), to monitor drought, in which BRF outperformed the other two models in terms of prediction. The method was later used by Nie et al. [18] to assess soil moisture, which is also one of the important factors affecting drought. Drought is a complex phenomenon influenced by various factors, and its non-linear characteristics make it challenging to predict accurately [19]. Neural networks do not rely on the mutual independence of variables; deep learning methods can effectively learn the complex features in drought and are an effective tool for drought prediction [20]. There has been a great deal of research showing that deep learning has gained good performance in the field of prediction [21,22]. Agana et al. [23] used the Deep Belief Network to make long-term predictions about drought, and they found that this method is better than the traditional MLP method and SVM method in terms of root mean square error (RMSE) and mean absolute error (MSE). Mokhtar et al. [24] used random forest (RF), extreme gradient boosting (XGB), the convolutional neural network (CNN), and long short-term memory (LSTM) to analyze drought on the eastern edge of the Qinghai–Tibet Plateau, obtaining favorable results. However, most drought predictions have focused on time series and ignored spatial scale impacts [25,26]. Droughts in a particular region are often influenced not only by local factors but also by climate conditions in distant areas. For instance, evaporation and precipitation are key components of the water cycle system, and some of the evaporated water can travel over long distances before it condenses into precipitation, affecting regions located hundreds of kilometers away [27,28]. Ham et al. [29] utilized the CNN for the long-term prediction of El Niño/Southern Oscillation (ENSO) and later optimized the method [30]. Their study aimed to predict Nino3.4, incorporating a broad range of spatial data during training to consider the influence of various regions on ENSO events.

As a complex natural disaster, it is difficult for a definition of drought to be unanimously accepted by the public due to the many factors affecting drought. Consequently, various drought indices have been developed, each with its own advantages and disadvantages [31]. The Palmer Drought Severity Index (PDSI) [32] is one of the classic drought indices, and many studies have used self-calibrating PDSI (scPDSI) for drought assessment [33]. There are also other drought indices, such as the standardized precipitation index (SPI) [34] and the standardized precipitation evapotranspiration index (SPEI) [35]. The SPEI is based on a water balance; related studies also showed that the SPEI can better reveal drought conditions in China [36]. Short-term drought prediction remains a challenging task [12], and early warnings for flash droughts are essential in the short term [37]. A new index called the Standardized Antecedent Precipitation Evapotranspiration Index (SAPEI) has been proposed; it utilizes precipitation and potential evapotranspiration and represents the surplus or deficit of surface water. The daily scale SAPEI helps authorities to make early and timely warnings [38].

Drought prediction is essential to effectively mitigate the impacts of drought [39]. However, existing drought prediction methods have generally performed well on monthly or longer time scales, while our study aimed to predict drought on a daily scale. Short-term drought is more difficult to predict than long-term drought, mainly because short-term meteorological and hydrological processes are relatively complex. Besides precipitation, temperature, and potential evapotranspiration, other variables will have a significant impact on short-term drought [12,40]. The prediction of drought in this study is not limited to the effect of a single meteorological factor on the time series; the method is a multivariate prediction that takes into account the effect of spatial extent. This also enhances the reliability of the findings. Deep learning methods can effectively learn the characteristics of different meteorological elements and greatly improve the prediction accuracy. The aim of our study is to make a daily prediction of the drought climate in a particular basin. The SAPEI data and an all-season CNN (A_CNN) model [30] are used for daily drought prediction at different time leads. Experiments are conducted in the Fenhe River (FHR) basin, which is a sub-basin in the middle of the HHH basin. The spatially averaged SAPEI of the FHR basin is predicted. More importantly, the spatially explicit meteorological conditions of the HHH basin are used as CNN model inputs to consider the impacts of surrounding regions. Here, we summarize the innovation of the research. Firstly, daily-scale drought prediction can provide timely and effective early warnings for droughts. Secondly, our experiment uses less training data to obtain better prediction results, and training on key variables affecting drought helps to improve the prediction accuracy. Thirdly, the drought prediction of the FHR basin has taken into account the influence of the surrounding environment, which makes the prediction more scientific and reliable.

## 2. Materials and Methods

### *2.1. Study Area*

The Fenhe River is the second largest tributary of the Yellow River, with a basin area of 39,741 km$^2$ [41]. It is an important ecological function area with high population density and a developed agricultural economy. It appears to be a highly drought-prone area in the warming climate. The HHH basin (95°~123° E, 30°~43° N) is located in the eastern part of China. It consists of three basins, namely the Yellow River Basin, the Huaihe River Basin, and the Haihe River Basin. The HHH basin covers an area of $1.433 \times 10^6$ km$^2$ and is a relatively developed economic region in China. However, climate change is more frequent in this region, and there is also a high incidence of meteorological disasters [42,43]. It is particularly vulnerable to extreme droughts with significant impacts [44–46].

This experiment used data from the HHH basin to train the model and then predicted the drought conditions in the FHR basin for the next ten days. Figure 1 shows the location of the study area.

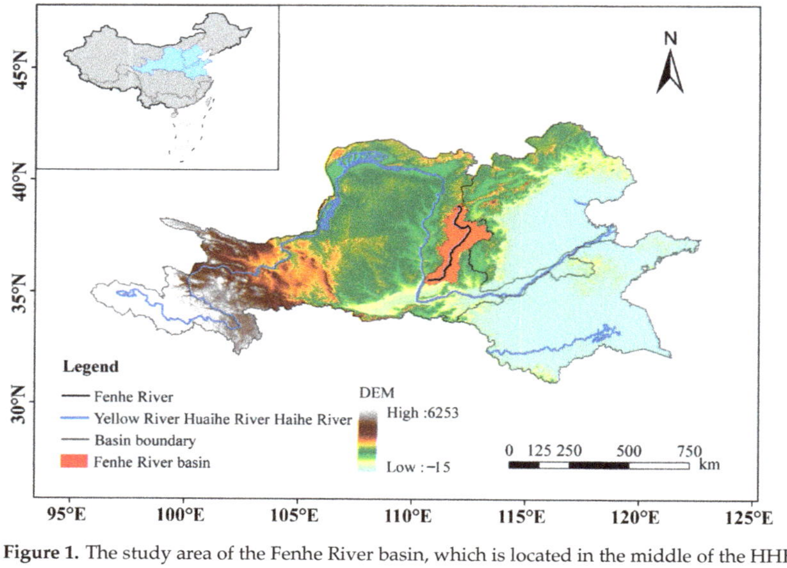


**Figure 1.** The study area of the Fenhe River basin, which is located in the middle of the HHH basin. The blue and black lines represent the river, and the red part is the FHR basin.

### 2.2. Datasets

Interpolated meteorological data are used in this study, and the variables are shown in Table 1. These data were developed by interpolating observations from more than 2400 ground-based meteorological stations in China [47]. Our study used data from 1961 to 2020; all variables were based on daily data with a spatial resolution of $0.5° \times 0.5°$. Since atmospheric stress on evapotranspiration is an essential factor of land drought, we further calculated daily potential evapotranspiration (PET) and vapor pressure deficit (VPD) from interpolated daily observations, which were then used as inputs of the model [35,38,48].

**Table 1.** Meteorological variables used in this study.

| Variables | Unit |
|---|---|
| precipitation (PRE) | mm |
| relative humidity | -- |
| wind velocity | m/s |
| mean temperature | °C |
| max temperature | °C |
| min temperature | °C |
| sunshine duration | h |
| potential evapotranspiration (PET) | mm |
| vapor pressure deficit (VPD) | kPa |
| SAPEI | -- |

We used the data from 1961 to 2000 as a training set and those from 2001 to 2010 as a validation set. And we used data from 2011 to 2020 to evaluate the model's performance. The ratio of the training set, validation set, and test set was 4:1:1. For network prediction, the data of the previous 30 days were used to forecast drought conditions in the FHR basin for the next 10 days, resulting in a total of 14,461 training samples, and both the validation set and the test set contained 3611 samples each. To accelerate the model convergence, we standardized the data pixel-wisely using Equation (1). ${X^*}_t$ indicates the value after standardization at time $t$.

$$ {X^*}_t = \frac{X_t - X^{min}}{X^{max} - X^{min}} \tag{1} $$

where $X_t$ denotes the grid value for a day at a certain latitude and longitude, $X^{min}$ denotes the minimum value on the corresponding latitude and longitude time series, and $X^{max}$ denotes the maximum value on the corresponding latitude and longitude time series.

### 2.3. Methods

#### 2.3.1. SAPEI Calculation

The daily SAPEI was used as a measure of drought in the FHR basin [38]. The calculation of the SAPEI requires the construction of the daily difference between precipitation (PRE) and PET, which can be estimated by Equation (2). The Penman–Monteith method was used to estimate PET, which has a more physical basis than other methods [49].

$$ D = \sum_{n=0}^{\mathrm{N}} a^{\mathrm{n}} (\mathrm{PRE} - \mathrm{PET})_n \tag{2} $$

$$ a^{\mathrm{N}} = c \tag{3} $$

In Equation (2), $D$ indicates the daily difference between PRE and PET, $a$ is the attenuation constant, N is the number of days ahead, and $c$ is the fraction of contribution of the last day of precipitation. Based on previous studies [38,50], $a = 0.98$ and $c = 13\%$, resulting in N = 100. After that, we could obtain the SAPEI based on the sequence $D$ [35,51].

The SAPEI values were divided into nine classes, namely extreme wet, severe wet, moderate wet, mild wet, normal, mild drought, moderate drought, severe drought, and extreme drought, as detailed in Table 2 [52]. We calculated the cumulative probability of the SAPEI in the test set in order to facilitate the evaluation of the forecast results at a later stage.

**Table 2.** Different categories of SAPEI.

| Threshold Value of SAPEI | Grade Name | Cumulative Probability (%) |
|---|---|---|
| SAPEI > 2.0 | Extremely wet | 0.64 |
| 1.5 < SAPEI ≤ 2.0 | Severe wet | 5.61 |
| 1.0 < SAPEI ≤ 1.5 | Moderate wet | 11.50 |
| 0.5 < SAPEI ≤ 1.0 | Mild wet | 20.50 |
| −0.5 < SAPEI ≤ 0.5 | Normal | 45.03 |
| −1.0 < SAPEI ≤ −0.5 | Mild drought | 11.81 |
| −1.5 < SAPEI ≤ −1.0 | Moderate drought | 4.13 |
| −2.0 < SAPEI ≤ −1.5 | Severe drought | 0.78 |
| SAPEI ≤ −2.0 | Extremely drought | 0 |

2.3.2. The A_CNN Model

The A_CNN model was used to predict the daily SAPEI [30]. This model combines the advantages of the CNN in spatial information processing, comprehensively considers the influence of the surrounding climate environment on the FHR basin, and learns the characteristics of continuous changes in drought in different seasons. It contains three convolutional layers. After convolutional layer1 and layer2, a max pooling layer is connected, and the three convolution layers are followed by two fully connected layers (as shown in Figure 2). We set a fixed time step p for the historical data $X = \{x_t, x_{t-1}, \ldots, x_{t-p}\}$ as input time series, where xt is the value of the variable at time t. The model output was a time series $Y$ with length $q$, $Y = \{x_{t+1}, x_{t+2}, \ldots, x_{t+q}\}$. For defining $p$ and $q$, we needed to pay attention to the periodic characteristics of the original data, so we set the time step to one month (30 days), and the output time of the model was 10 days. In our study, the input data were the spatio-temporal cubes of two sets of variables at time $t-30$ to $t-1$, and the output was the average value of the SAPEI index of the FHR basin from $t+0$ to $t+10$ days. M denotes the number of convolutional kernels (the value is 10), and N denotes the number of nodes in the FC layer (the value is 50). It is worth noting that our inputs were multiple spatio-temporal datasets with a spatial range of (95°~123° E, 30°~43° N), which was different from the traditional multivariate time series forecasting. The SAPEI index of the FHR basin was spatially averaged to show the basin-scale drought, which was used as the output of the model.

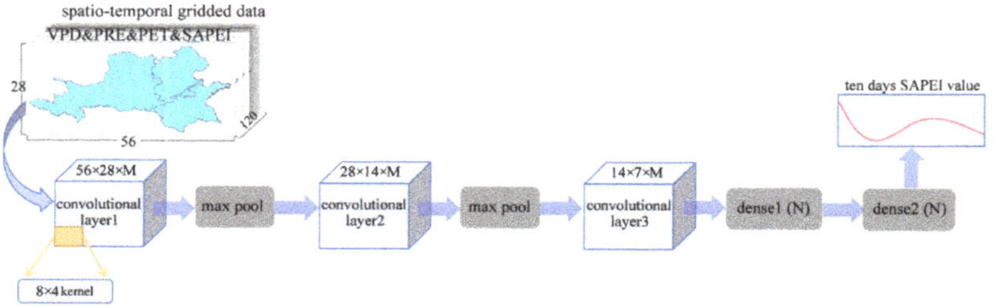


**Figure 2.** The core structure of the model consists of three convolutional layers, two max-pooling layers, two fully connected layers, and an output layer. Model components are differentiated by color and the red wavy line indicates the SAPEI value.

In the training process, the convolutional kernel size was set to 8 × 4 in the first convolutional layer and to 4 × 2 in the last two convolutional layers. The number of epochs was 100, with each epoch containing 56 iterations, and the Adam optimizer was used. The hyperbolic tangent function (tanh) was used for the activation function, and the learning rate of training was fixed at 0.0005. The output shapes of each layer in the model are summarized in Table 3.

**Table 3.** Output shape for each layer of the model.

| Layer | Output Shape |
|---|---|
| input layer | 56 × 28 × 120 |
| convolutional layer1 | 56 × 28 × 10 |
| max pool | 28 × 14 × 10 |
| convolutional layer2 | 28 × 14 × 10 |
| max pool | 14 × 7 × 10 |
| convolutional layer3 | 14 × 7 × 10 |
| dense1 | 50 |
| dense2 | 50 |
| output layer | 10 |

PRE, PET, and VPD play a crucial role in the water cycle and have significant impacts on the variability in drought levels in a region [53–55]. VPD is mainly expressed as the difference between the water vapor pressure at saturation and the actual water vapor pressure at a certain temperature [48,56]. To improve the prediction accuracy and speed up the convergence of the model, we designed two sets of experiments for comparison. In the first experiment, the SAPEI and three important variables affecting drought (PRE, PET, VPD) were used as input variables, which we named EXP4. In the second experiment, we took all meteorological variables as input (as shown in Table 1), which we named EXP10. We wanted to explore whether we could improve the prediction accuracy by capturing the key factors affecting drought.

#### 2.3.3. Evaluation Metrics

To evaluate the model performance, the mean squared error (MSE) was used as the loss function; the mean absolute error (MAE), BIAS, Nash–Sutcliffe efficiency (NSE), Kling–Gupta efficiency (KGE), and the Pearson correlation coefficient (R) were also calculated for the test set. To minimize the impact of RMSE deviations, we employed the unbiased root mean square difference (ubRMSD) metric, which combines R and standard deviation (Std) to evaluate the effect of EXP4 and EXP10. The calculation of the evaluation metrics is summarized as follows:

$$\mathrm{MSE}(y,y') = \frac{\sum_{i=1}^{n} (y_i - y_i')^2}{n} \tag{4}$$

$$\mathrm{MAE}(y,y') = \frac{1}{n}\sum_{i=1}^{n} |y_i - y_i'| \tag{5}$$

$$\mathrm{BIAS} = \frac{\sum_{i=1}^{n} (y_i' - y_i)}{n} \tag{6}$$

$$\mathrm{R}(y,y') = \frac{Cov(y,y')}{\sqrt{Var(y)Var(y')}} \tag{7}$$

$$\mathrm{NSE}(y,y') = 1 - \frac{\sum_{i=1}^{n} (y_i' - y_i)^2}{\sum_{i=1}^{n} (y_i - \overline{y})^2} \tag{8}$$

$$\mathrm{KGE}(y,y') = 1 - \sqrt{(\mathrm{R}-1)^2 + (\alpha-1)^2 + (\beta-1)^2} \tag{9}$$

$$\mathrm{RMSD} = \sqrt{\mathrm{MSE}} \tag{10}$$

$$\mathrm{ubRMSD} = \sqrt{\mathrm{RMSD}^2 - \mathrm{BIAS}^2} \tag{11}$$

In Equations (4)–(9), $y_i$ denotes the value of SAPEI on day $i$, $y_i'$ denotes the predicted value, $\alpha$ denotes the ratio between the Std of predicted results and the Std of true values, and $\beta$ denotes the ratio between the mean of predicted results and the mean of true values. $Var(y)$ and $Var(y')$ mean the variance in the observations and predictions, and $Cov(y,y')$ is their covariance.

## 3. Results

### 3.1. Identifying Key Variables to Enhance Prediction Accuracy

In this experiment, the SAPEI was predicted for the next 10 days in the FHR basin. Non-meteorological parameters such as the SAPEI were calculated in our study; we removed non-meteorological parameters and trained the model, which we named EXP7. Table 4 shows the prediction results for all parameters (EXP10) and the removal of non-meteorological parameters (EXP7). We show the predictions for days 1, 3, 5, 7, and 9, and the five prediction lengths are denoted as PL1, PL3, PL5, PL7, and PL9. When we remove the non-meteorological parameters, the model's prediction results significantly deteriorate and become inaccurate. A higher R and lower MSE indicate a better prediction effect. The R value for the predicted results of EXP7, as shown in Table 4, is significantly lower than that of EXP10, while the MSE is noticeably higher than that of EXP10. Non-meteorological parameters play a crucial role in the prediction process. Therefore, we chose to include non-meteorological parameters in both sets of comparative experiments.

**Table 4.** Comparison of results for different prediction lengths.

| | | PL1 | PL3 | PL5 | PL7 | PL9 |
|---|---|---|---|---|---|---|
| EXP7 | MSE | 0.413 | 0.498 | 0.503 | 0.703 | 0.648 |
| | R | 0.695 | 0.693 | 0.673 | 0.631 | 0.588 |
| EXP10 | MSE | 0.100 | 0.193 | 0.273 | 0.310 | 0.411 |
| | R | 0.926 | 0.883 | 0.838 | 0.795 | 0.740 |

We compared the predictions on day 1, 3, 5, 7, and 9 from two experiments. A Taylor diagram is used to compare the evaluation metrics between predicted and actual values. Figure 3 displays the training results obtained using two experiments for different prediction lengths (indicated by colors). The R values range from 0.8 to 0.96, indicating that both experiments have achieved good results for different prediction lengths. However, upon comparison, it was concluded that the results obtained by training with EXP4 were significantly better than EXP10. This is evident in the distribution of the squares, which is consistently below the circles of the same color, indicating a higher correlation than the results obtained by training with all variables. In addition, the ubRMSD of EXP4 is significantly lower than EXP10, which shows that for different prediction lengths, EXP4 gives better results than EXP10. The Std in the Taylor diagram is the ratio value between the Std of predicted data from the model and true values. The ratio closer to 1 indicates the better prediction of results. As can be seen from the Taylor diagram, for a prediction length of 3, EXP10 is slightly better than EXP4. However, EXP4 outperforms EXP10 in the rest of the predicted lengths. Combining the three evaluation metrics, EXP4 achieves a better performance. To verify this conclusion, the test set samples of the two experiments were stitched together, and the MSE was calculated for predicting day one to day ten, respectively. The mean of the ten-day prediction MSE reached 0.169 for the EXP4 and only 0.272 for the mean of the EXP10, which further confirms the previous conclusion.

While deep learning methods often require a substantial amount of data to produce optimal results, the size of the dataset does not always guarantee improved performance. The properties of the data and the model itself can significantly impact the training outcome, and it is essential to evaluate them within the context of the specific experiment. In our experiment, we obtained favorable outcomes for EXP4. By focusing on the important variables, the model could more easily discern the time series' characteristics. Furthermore, increasing the data quantity did not improve the results, so we mainly present the EXP4 findings in the following section.

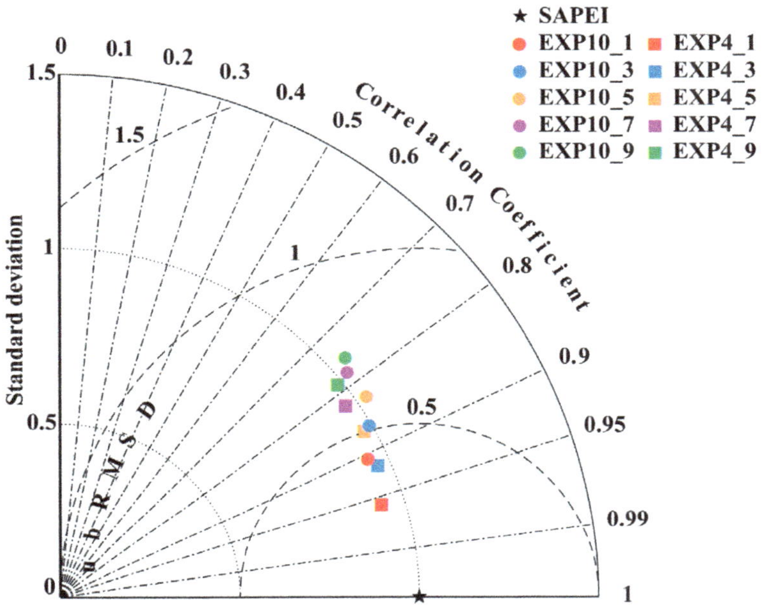


**Figure 3.** Comparison of the results of EXP4 and EXP10: We record the predictions for the two sets of experiments one day ahead as EXP4_1 and EXP10_1, and so on. The Taylor diagram shows the predictive effect of different models in considering Standard deviation (dotted), Correlation Coefficient (alternating lines and dots), and ubRMSD (dashed).

The MAE is commonly used to assess the deviation between the predicted and actual SAPEI values; it is calculated in a way that avoids the influence of some extreme values on the overall results, making the results more stable [57–59]. Figure 4 shows that the MAE of the test set gradually rises as the prediction length increases. The MAE for each prediction length is shown as a red line, and the shaded area indicates the standard deviation. Although the effect becomes progressively worse as the prediction length increases, the MAE for the next ten-day forecast of the FHR basin remains stable below 0.4, indicating that the model's performance is generally relatively consistent.

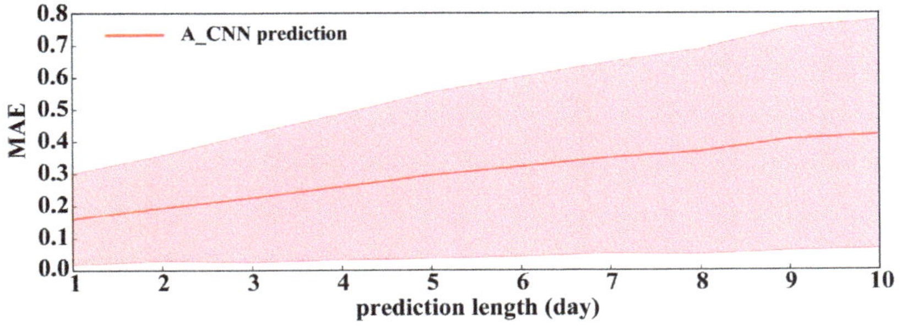


**Figure 4.** The MAE from 1- to 10-day prediction, and the shaded area indicates the standard deviation.

This study also applied the NSE and KGE to evaluate the model's predictions; Table 5 shows the different evaluation metrics of the model in the test set. It can be seen from Table 5 that the model's predictions are more accurate when the prediction length is 1 day (PL1) or 3 days (PL3). However, the model's predictions deteriorate when the prediction length increases. When the prediction length changes to 9 days (PL9), the R of the predicted results is acceptable, and in terms of the KGE, the model does not give accurate predictions.

**Table 5.** Evaluation metrics for models with different prediction lengths.

| Prediction Length | NSE | KGE | MSE | R |
|---|---|---|---|---|
| PL1 | 0.922 | 0.875 | 0.046 | 0.961 |
| PL3 | 0.845 | 0.757 | 0.091 | 0.923 |
| PL5 | 0.736 | 0.522 | 0.154 | 0.877 |
| PL7 | 0.637 | 0.438 | 0.212 | 0.830 |
| PL9 | 0.508 | 0.122 | 0.286 | 0.793 |

In order to provide a more intuitive representation of the model's prediction performance, we extracted the prediction results on days 1, 3, 5, 7, and 9 for the test set samples and plotted time series graphs for comparison. Figure 5 displays the time series of predicted values and true values from 2011 to 2020. The true value is represented by the black line, which is the SAPEI calculated from meteorological data. The other five colors correspond to the five different prediction lengths. For each predicted time series, we calculated the R and MSE. The model performed the best in predicting the next day, with an R of 0.96. As the prediction time increased, the effect of model deteriorated slightly, but the overall performance remained relatively stable. Severe drought events were defined as SAPEI values below −1.5, which are represented by a black dashed line in the figure. Although the frequency of severe drought events in the FHR basin was low over the past decade, the model accurately predicted them. Additionally, the model showed promising performance in predicting gradual climate changes from wet to drought conditions.

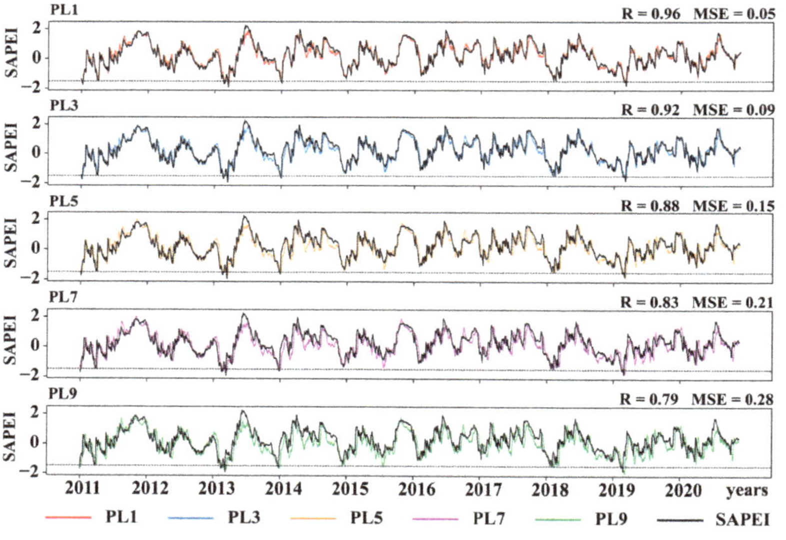


**Figure 5.** The true and predicted SAPEI values; five different prediction lengths were selected for presentation. The black line indicates true SAPEI values, and the other colored lines indicate 1-day, 3-day, 5-day, 7-day, and 9-day predictions, respectively. The dotted line indicates a SAPEI value of −1.5, beyond which, it is a severe drought.

Then, the prediction performances of different SAPEI categories as listed in Table 2 were evaluated using the ten-year test dataset spanning from 2011 to 2020 (as shown in Figure 6). Clearly, the prediction bias increased for SAPEI extremes with increasing prediction lengths. From a model training perspective, this is not unexpected since the SAPEI has fewer extreme samples, making it difficult for the model to learn their features, leading to relatively poor prediction skills. It is striking to find that the used model overestimated the drought conditions and underestimated the wet conditions. The prediction biases of drought events are relatively smaller than those of wet events, indicating the better performance of the used model for drought prediction.

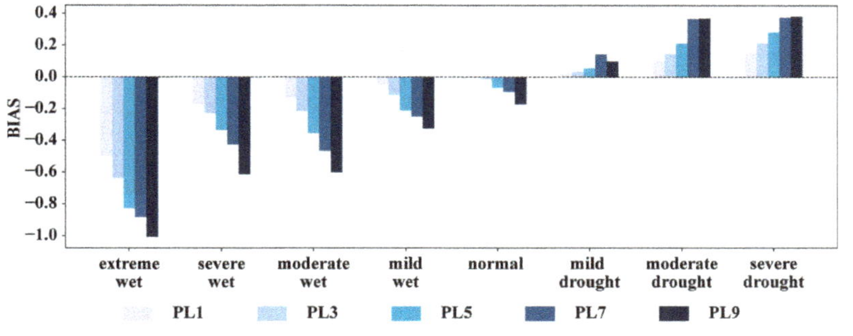


**Figure 6.** The prediction BIAS of each SAPEI category at different prediction lengths.

### 3.2. *Predicting Severe Drought*

We used a case of severe drought to evaluate the model performance. In early 2019, there was high temperature with significant precipitation deficit in the HHH region, leading to a severe drought. Such a drought event is reflected in the SAPEI data, and thus, we used it for the case study. In Figure 7a, 1-day, 3-day, 5-day, 7-day, and 9-day predictions against the realistic SAPEI values are shown. There was significant precipitation deficit in March, and PET was increasing due to high temperature, which caused a decreasing SAPEI and thus a severe drought. In April, several precipitation events alleviated the drought. It appears that the basin-scale severe drought event was well predicted by the used model, especially when short prediction lengths were used, suggesting the model's credibility in the prediction of extreme events.

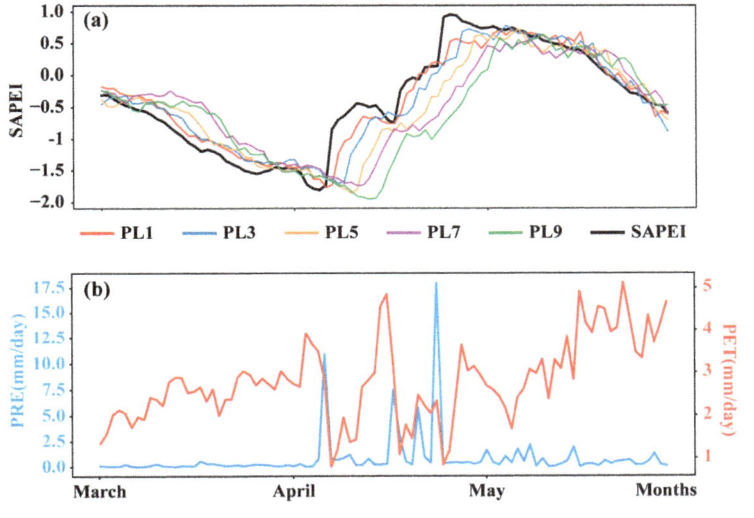


**Figure 7.** (**a**) The SAPEI predictions in early 2019 in the FHR basin. The black line represents the real SAPEI value, and the other colored lines represent the predictions at different lengths. (**b**) The PRE and PET.

## 4. Discussion

In this study, the SAEPI was selected as a drought index to assess drought conditions in the FHR basin. We used a deep learning approach to predict the next 10 days of drought in the FHR basin. Overall, the predictions of the model are accurate. As shown in Figure 5, the model's R reaches 0.79 even when the prediction length is 9 days. However, the accuracy of model predictions inevitably decreases as the length of the prediction increases; this is also consistent with previous research [60]. The higher prediction accuracy obtained by using fewer key variables for training also reflects that deep learning can overcome the limitations of numerical weather prediction [61,62]. Xu et al. [26] have demonstrated that deep learning methods can efficiently process time series, while our study takes spatial factors into account, making the prediction of drought more scientific. The training data for the model contain non-meteorological parameters, which also has some limitations. Non-meteorological parameters enhance prediction accuracy, but they are calculated from meteorological parameters and are not independent.

Although the experiment used these as key variables affecting drought, it does not mean that the other variables have little influence on drought. Temperature changes can also have a significant impact on drought conditions [63]. We also tried to add variables such as temperature and wind speed for training; the results were stable, but the optimal solution was not obtained. We need to note that a single increase in predictors does not necessarily improve model performance [19]. Dikshit et al. [64] mentioned that exploring the effects of different meteorological elements in drought prediction can improve the accuracy of predictions. The purpose of this study is to make timely and effective predictions of short-term drought; this contributes to a timely response by policy makers, farmers, and other stakeholders [65].

In our study, we calculated mean values for the SAPEI within the FHR basin to produce labels for our model. While this approach provides an overview of the drought situation in the study area, it also presents a challenge of weakening extreme values that could significantly impact the overall analysis. In addition, predictor selection plays an important role in drought prediction [25]. Our study takes into account the influence of the spatial environment; in addition to natural factors, the impact of human activities on climate change should not be ignored. Combining the impact of multiple factors will allow us to refine our model, contributing to the development of better disaster prevention strategies.

## 5. Conclusions

Our study utilized a deep learning method to predict drought in the FHR basin on a daily scale. Our methodology takes into account the effects of multiple meteorological elements and spatial scales on drought. The results of the study show that the prediction accuracy of the model decreases with increasing prediction lengths. As can be seen in Table 4, the NSE of the prediction results reaches 0.922 when the prediction length is 1 day, but the accuracy of the model decreases when the prediction length increases. The long-term prediction of the model did not reach a high level of confidence in terms of the KGE. However, previous studies have also demonstrated that this is an acceptable phenomenon.

It is concluded from our study that for predicting drought in the FHR basin, EXP4, which used the important variables affecting drought as network inputs, obtained a better result compared to EXP10. We used relatively little data to obtain better predictions. Capturing the characteristics of several important variables that affect drought allows the model to make more effective predictions. This reduces computational costs significantly compared to traditional numerical weather prediction methods [62].

The model is able to capture fluctuations in SAPEI and predict heavy drought events in advance, which effectively mitigates the losses caused by natural disasters. Timely access to drought-related early warning information is key for early warning institutions to select adaptation strategies [66]. Therefore, this research has important implications in drought prevention.

**Author Contributions:** Conceptualization, Z.C., G.W. and X.W.; methodology, Z.C. and X.W.; software, Z.C.; validation, Z.C. and G.W.; formal analysis, Z.C. and Y.H.; resources, H.J.; data curation, Z.C.; writing—original draft preparation, Z.C.; writing—review and editing, G.W. and Z.D.; visualization, Z.C.; supervision, G.W., X.W. and Y.L.; project administration, G.W.; funding acquisition, G.W. All authors have read and agreed to the published version of the manuscript.

**Funding:** This research was funded by the National Natural Science Foundation of China, grant number 42275028.

**Institutional Review Board Statement:** Not applicable.

**Informed Consent Statement:** Not applicable.

**Data Availability Statement:** Data are contained within the article.

**Acknowledgments:** We thank the researchers or teams who provided the basic data. We also thank the authors, reviewers, and editors who made amendments to the article.

**Conflicts of Interest:** The authors declare no conflicts of interest.

## References

1. Dikshit, A.; Pradhan, B.; Huete, A.; Park, H.-J. Spatial based drought assessment: Where are we heading? A review on the current status and future. *Sci. Total Environ.* **2022**, *844*, 157239. [CrossRef] [PubMed]
2. Shi, H.; Zhou, Z.; Liu, L.; Liu, S. A global perspective on propagation from meteorological drought to hydrological drought during 1902–2014. *Atmos. Res.* **2022**, *280*, 106441. [CrossRef]
3. Jehanzaib, M.; Yoo, J.; Kwon, H.-H.; Kim, T.-W. Reassessing the frequency and severity of meteorological drought considering non-stationarity and copula-based bivariate probability. *J. Hydrol.* **2021**, *603*, 126948. [CrossRef]
4. Gavahi, K.; Abbaszadeh, P.; Moradkhani, H. How does precipitation data influence the land surface data assimilation for drought monitoring? *Sci. Total Environ.* **2022**, *831*, 154916. [CrossRef] [PubMed]
5. Wang, C.; Chen, J.; Gu, L.; Wu, G.; Tong, S.; Xiong, L.; Xu, C.-Y. A pathway analysis method for quantifying the contributions of precipitation and potential evapotranspiration anomalies to soil moisture drought. *J. Hydrol.* **2023**, *621*, 129570. [CrossRef]
6. Han, L.; Zhang, Q.; Zhang, Z.; Jia, J.; Wang, Y.; Huang, T.; Cheng, Y. Drought area, intensity and frequency changes in China under climate warming, 1961–2014. *J. Arid Environ.* **2021**, *193*, 104596. [CrossRef]
7. Xu, Y.; Zhu, X.; Cheng, X.; Gun, Z.; Lin, J.; Zhao, J.; Yao, L.; Zhou, C. Drought assessment of China in 2002–2017 based on a comprehensive drought index. *Agric. For. Meteorol.* **2022**, *319*, 108922. [CrossRef]
8. Su, B.; Huang, J.; Fischer, T.; Wang, Y.; Kundzewicz, Z.W.; Zhai, J.; Sun, H.; Wang, A.; Zeng, X.; Wang, G. Drought losses in China might double between the 1.5 C and 2.0 C warming. *Proc. Natl. Acad. Sci. USA* **2018**, *115*, 10600–10605. [CrossRef]
9. Wang, Z.; Li, J.; Lai, C.; Zeng, Z.; Zhong, R.; Chen, X.; Zhou, X.; Wang, M. Does drought in China show a significant decreasing trend from 1961 to 2009? *Sci. Total Environ.* **2017**, *579*, 314–324. [CrossRef]
10. Shi, X.; Ding, H.; Wu, M.; Zhang, N.; Shi, M.; Chen, F.; Li, Y. Effects of different types of drought on vegetation in Huang-Huai-Hai River Basin, China. *Ecol. Indic.* **2022**, *144*, 109428. [CrossRef]
11. Zhan, C.; Liang, C.; Zhao, L.; Jiang, S.; Niu, K.; Zhang, Y. Drought-related cumulative and time-lag effects on vegetation dynamics across the Yellow River Basin, China. *Ecol. Indic.* **2022**, *143*, 109409. [CrossRef]
12. Zhang, X.; Duan, Y.; Duan, J.; Chen, L.; Jian, D.; Lv, M.; Yang, Q.; Ma, Z. A daily drought index-based regional drought forecasting using the Global Forecast System model outputs over China. *Atmos. Res.* **2022**, *273*, 106166. [CrossRef]
13. Balti, H.; Abbes, A.B.; Mellouli, N.; Farah, I.R.; Sang, Y.; Lamolle, M. A review of drought monitoring with big data: Issues, methods, challenges and research directions. *Ecol. Inf.* **2020**, *60*, 101136. [CrossRef]
14. Durdu, Ö.F. Application of linear stochastic models for drought forecasting in the Büyük Menderes river basin, western Turkey. *Stoch. Environ. Res. Risk Assess.* **2010**, *24*, 1145–1162. [CrossRef]
15. Li, H.; Sun, B.; Wang, H.; Zhou, B.; Duan, M. Mechanisms and physical-empirical prediction model of concurrent heatwaves and droughts in July–August over northeastern China. *J. Hydrol.* **2022**, *614*, 128535. [CrossRef]
16. Prodhan, F.A.; Zhang, J.; Hasan, S.S.; Sharma, T.P.P.; Mohana, H.P. A review of machine learning methods for drought hazard monitoring and forecasting: Current research trends, challenges, and future research directions. *Environ. Model. Softw.* **2022**, *149*, 105327. [CrossRef]
17. Feng, P.; Wang, B.; Li Liu, D.; Yu, Q. Machine learning-based integration of remotely-sensed drought factors can improve the estimation of agricultural drought in South-Eastern Australia. *Agric. Syst.* **2019**, *173*, 303–316. [CrossRef]
18. Nie, H.; Yang, L.; Li, X.; Ren, L.; Xu, J.; Feng, Y. Spatial prediction of soil moisture content in winter wheat based on machine learning model. In Proceedings of the 2018 26th International Conference on Geoinformatics, Kunming, China, 28–30 June 2018; pp. 1–6.
19. Dikshit, A.; Pradhan, B.; Santosh, M. Artificial neural networks in drought prediction in the 21st century–A scientometric analysis. *Appl. Soft Comput.* **2022**, *114*, 108080. [CrossRef]

20. Nourani, V.; Baghanam, A.H.; Adamowski, J.; Kisi, O. Applications of hybrid wavelet–artificial intelligence models in hydrology: A review. *J. Hydrol.* **2014**, *514*, 358–377. [CrossRef]
21. Coşkun, Ö.; Citakoglu, H. Prediction of the standardized precipitation index based on the long short-term memory and empirical mode decomposition-extreme learning machine models: The Case of Sakarya, Türkiye. *Phys. Chem. Earth Parts A/B/C* **2023**, *131*, 103418. [CrossRef]
22. Zhang, X.; Jin, Q.; Yu, T.; Xiang, S.; Kuang, Q.; Prinet, V.; Pan, C. Multi-modal spatio-temporal meteorological forecasting with deep neural network. *ISPRS J. Photogramm. Remote Sens.* **2022**, *188*, 380–393. [CrossRef]
23. Agana, N.A.; Homaifar, A. A deep learning based approach for long-term drought prediction. In Proceedings of the SoutheastCon 2017, Concord (Charlotte), NC, USA, 30 March–2 April 2017; pp. 1–8.
24. Mokhtar, A.; Jalali, M.; He, H.; Al-Ansari, N.; Elbeltagi, A.; Alsafadi, K.; Abdo, H.G.; Sammen, S.S.; Gyasi-Agyei, Y.; Rodrigo-Comino, J. Estimation of SPEI meteorological drought using machine learning algorithms. *IEEE Access* **2021**, *9*, 65503–65523. [CrossRef]
25. Mei, P.; Liu, J.; Liu, C.; Liu, J. A deep learning model and its application to predict the monthly MCI drought index in the Yunnan Province of China. *Atmosphere* **2022**, *13*, 1951. [CrossRef]
26. Xu, Z.; Sun, H.; Zhang, T.; Xu, H.; Wu, D.; Gao, J. Evaluating established deep learning methods in constructing integrated remote sensing drought index: A case study in China. *Agric. Water Manag.* **2023**, *286*, 108405. [CrossRef]
27. Dirmeyer, P.A.; Brubaker, K.L. Characterization of the global hydrologic cycle from a back-trajectory analysis of atmospheric water vapor. *J. Hydrometeorol.* **2007**, *8*, 20–37. [CrossRef]
28. Wei, J.; Dirmeyer, P.A. Sensitivity of land precipitation to surface evapotranspiration: A nonlocal perspective based on water vapor transport. *Geophys. Res. Lett.* **2019**, *46*, 12588–12597. [CrossRef]
29. Ham, Y.-G.; Kim, J.-H.; Luo, J.-J. Deep learning for multi-year ENSO forecasts. *Nature* **2019**, *573*, 568–572. [CrossRef]
30. Ham, Y.-G.; Kim, J.-H.; Kim, E.-S.; On, K.-W. Unified deep learning model for El Niño/Southern Oscillation forecasts by incorporating seasonality in climate data. *Sci. Bull.* **2021**, *66*, 1358–1366. [CrossRef]
31. Hao, Z.; Singh, V.P. Drought characterization from a multivariate perspective: A review. *J. Hydrol.* **2015**, *527*, 668–678. [CrossRef]
32. Palmer, W.C. *Meteorological Drought*; US Department of Commerce, Weather Bureau: Washington, DC, USA, 1965; Volume 30.
33. Wells, N.; Goddard, S.; Hayes, M.J. A self-calibrating Palmer drought severity index. *J. Clim.* **2004**, *17*, 2335–2351. [CrossRef]
34. McKee, T.B.; Doesken, N.J.; Kleist, J. The relationship of drought frequency and duration to time scales. In Proceedings of the 8th Conference on Applied Climatology, Anaheim, CA, USA, 17–22 January 1993; pp. 179–183.
35. Vicente-Serrano, S.M.; Beguería, S.; López-Moreno, J.I. A multiscalar drought index sensitive to global warming: The standardized precipitation evapotranspiration index. *J. Clim.* **2010**, *23*, 1696–1718. [CrossRef]
36. Wang, L.; Chen, W. Applicability analysis of standardized precipitation evapotranspiration index in drought monitoring in China. *Plateau Meteorol.* **2014**, *33*, 423–431.
37. Yuan, X.; Wang, Y.; Ji, P.; Wu, P.; Sheffield, J.; Otkin, J.A. A global transition to flash droughts under climate change. *Science* **2023**, *380*, 187–191. [CrossRef]
38. Li, J.; Wang, Z.; Wu, X.; Xu, C.-Y.; Guo, S.; Chen, X. Toward monitoring short-term droughts using a novel daily scale, standardized antecedent precipitation evapotranspiration index. *J. Hydrometeorol.* **2020**, *21*, 891–908. [CrossRef]
39. Hao, Z.; Singh, V.P.; Xia, Y. Seasonal drought prediction: Advances, challenges, and future prospects. *Rev. Geophys.* **2018**, *56*, 108–141. [CrossRef]
40. Park, S.; Seo, E.; Kang, D.; Im, J.; Lee, M.-I. Prediction of drought on pentad scale using remote sensing data and MJO index through random forest over East Asia. *Remote Sens.* **2018**, *10*, 1811. [CrossRef]
41. Wu, B.; Yang, S.; Shao, N.; Peng, R.; Guan, Y. Effects of land use change on ecosystem service value in fragile ecological area of the Loess Plateau a case study of Fenhe River Basin. *Soil Water Conserv.* **2019**, *26*, 340–345.
42. Yan, D.; Wu, D.; Huang, R.; Wang, L.; Yang, G. Drought evolution characteristics and precipitation intensity changes during alternating dry–wet changes in the Huang–Huai–Hai River basin. *Hydrol. Earth Syst. Sci.* **2013**, *17*, 2859–2871. [CrossRef]
43. Yuan, Y.; Yan, D.; Yuan, Z.; Yin, J.; Zhao, Z. Spatial distribution of precipitation in huang-huai-Hai River basin between 1961 to 2016, China. *Int. J. Environ. Res. Public Health* **2019**, *16*, 3404. [CrossRef]
44. Hu, W.; She, D.; Xia, J.; He, B.; Hu, C. Dominant patterns of dryness/wetness variability in the Huang-Huai-Hai River Basin and its relationship with multiscale climate oscillations. *Atmos. Res.* **2021**, *247*, 105148. [CrossRef]
45. Deng, Y.; Wu, D.; Wang, X.; Xie, Z. Responding time scales of vegetation production to extreme droughts over China. *Ecol. Indic.* **2022**, *136*, 108630. [CrossRef]
46. Li, C.; Kattel, G.R.; Zhang, J.; Shang, Y.; Gnyawali, K.R.; Zhang, F.; Miao, L. Slightly enhanced drought in the Yellow River Basin under future warming scenarios. *Atmos. Res.* **2022**, *280*, 106423. [CrossRef]
47. Wu, J.; Gao, X.-J. A gridded daily observation dataset over China region and comparison with the other datasets. *Chin. J. Geophys.* **2013**, *56*, 1102–1111.
48. Yuan, W.; Zheng, Y.; Piao, S.; Ciais, P.; Lombardozzi, D.; Wang, Y.; Ryu, Y.; Chen, G.; Dong, W.; Hu, Z. Increased atmospheric vapor pressure deficit reduces global vegetation growth. *Sci. Adv.* **2019**, *5*, eaax1396. [CrossRef]
49. Allen, R.G.; Pereira, L.S.; Raes, D.; Smith, M. Crop evapotranspiration-Guidelines for computing crop water requirements-FAO Irrigation and drainage paper 56. *Fao Rome* **1998**, *300*, D05109.
50. Heggen, R.J. Normalized antecedent precipitation index. *J. Hydrol. Eng.* **2001**, *6*, 377–381. [CrossRef]

51. Chen, J.; Yu, W.; Liu, R.; Yue, W.; Chen, X. Daily standardized antecedent precipitation evapotranspiration index (SAPEI) and its adaptability in Anhui Province. *Chin. J. Eco-Agric.* **2019**, *27*, 919–928.
52. Miah, M.G.; Abdullah, H.M.; Jeong, C. Exploring standardized precipitation evapotranspiration index for drought assessment in Bangladesh. *Environ. Monit. Assess.* **2017**, *189*, 547. [CrossRef]
53. Ma, T.; Liang, Y.; Lau, M.K.; Liu, B.; Wu, M.M.; He, H.S. Quantifying the relative importance of potential evapotranspiration and timescale selection in assessing extreme drought frequency in conterminous China. *Atmos. Res.* **2021**, *263*, 105797. [CrossRef]
54. Cook, B.I.; Smerdon, J.E.; Seager, R.; Coats, S. Global warming and 21st century drying. *Clim. Dyn.* **2014**, *43*, 2607–2627. [CrossRef]
55. Oogathoo, S.; Houle, D.; Duchesne, L.; Kneeshaw, D. Vapour pressure deficit and solar radiation are the major drivers of transpiration of balsam fir and black spruce tree species in humid boreal regions, even during a short-term drought. *Agric. For. Meteorol.* **2020**, *291*, 108063. [CrossRef]
56. Li, M.; Yao, J.; Guan, J.; Zheng, J. Observed changes in vapor pressure deficit suggest a systematic drying of the atmosphere in Xinjiang of China. *Atmos. Res.* **2021**, *248*, 105199. [CrossRef]
57. Legates, D.R.; McCabe, G.J., Jr. Evaluating the use of "goodness-of-fit" measures in hydrologic and hydroclimatic model validation. *Water Resour. Res.* **1999**, *35*, 233–241. [CrossRef]
58. Jayasinghe, W.L.P.; Deo, R.C.; Ghahramani, A.; Ghimire, S.; Raj, N. Development and evaluation of hybrid deep learning long short-term memory network model for pan evaporation estimation trained with satellite and ground-based data. *J. Hydrol.* **2022**, *607*, 127534. [CrossRef]
59. Ayris, D.; Imtiaz, M.; Horbury, K.; Williams, B.; Blackney, M.; See, C.S.H.; Shah, S.A.A. Novel deep learning approach to model and predict the spread of COVID-19. *Intell. Syst. Appl.* **2022**, *14*, 200068. [CrossRef]
60. Dai, R.; Wang, W.; Zhang, R.; Yu, L. Multimodal deep learning water level forecasting model for multiscale drought alert in Feiyun River basin. *Expert Syst. Appl.* **2023**, *244*, 122951. [CrossRef]
61. Abdulla, N.; Demirci, M.; Ozdemir, S. Design and evaluation of adaptive deep learning models for weather forecasting. *Eng. Appl. Artif. Intell.* **2022**, *116*, 105440. [CrossRef]
62. Ren, X.; Li, X.; Ren, K.; Song, J.; Xu, Z.; Deng, K.; Wang, X. Deep learning-based weather prediction: A survey. *Big Data Res.* **2021**, *23*, 100178. [CrossRef]
63. Hu, J.; Yang, Z.; Hou, C.; Ouyang, W. Compound risk dynamics of drought by extreme precipitation and temperature events in a semi-arid watershed. *Atmos. Res.* **2023**, *281*, 106474. [CrossRef]
64. Dikshit, A.; Pradhan, B. Interpretable and explainable AI (XAI) model for spatial drought prediction. *Sci. Total Environ.* **2021**, *801*, 149797. [CrossRef]
65. Otkin, J.A.; Woloszyn, M.; Wang, H.; Svoboda, M.; Skumanich, M.; Pulwarty, R.; Lisonbee, J.; Hoell, A.; Hobbins, M.; Haigh, T. Getting ahead of flash drought: From early warning to early action. *Bull. Am. Meteorol. Soc.* **2022**, *103*, E2188–E2202. [CrossRef]
66. Ndiritu, S.W. Drought responses and adaptation strategies to climate change by pastoralists in the semi-arid area, Laikipia County, Kenya. *Mitig. Adapt. Strategies Glob. Chang.* **2021**, *26*, 10. [CrossRef]

atmosphere

MDPI

*Article*

# Teleconnections of Atmospheric Circulations to Meteorological Drought in the Lancang-Mekong River Basin

**Lei Fan [1], Yi Wang [1,*], Chenglin Cao [1] and Wen Chen [2,3]**

1 School of Water Resources and Hydropower Engineering, North China Electric Power University, Beijing 102206, China; fanlei@ncepu.edu.cn (L.F.); ccl1209@ncepu.edu.cn (C.C.)
2 Key Laboratory of Watershed Geographic Sciences Nanjing, Institute of Geography and Limnology, Chinese Academy of Sciences, Nanjing 210008, China; wchen@niglas.ac.cn
3 Key Laboratory of Coastal Zone Exploitation and Protection, Ministry of Natural Resources, Nanjing 210024, China
* Correspondence: wangyi28@ncepu.edu.cn

**Abstract:** The Lancang-Mekong River Basin (LMRB) is one of the major transboundary basins globally, facing ongoing challenges due to flood and drought disasters. Particularly in the past two decades, the basin has experienced an increased frequency of meteorological drought events, posing serious threats to the local socio-economic structures and ecological systems. Thus, this study aimed to analyze the meteorological drought characteristics in the LMRB and identify the impact and correlation of atmospheric circulation on the meteorological drought in the basin. Specifically, the different levels of meteorological drought events were defined using the Run Theory based on the seasonal and annual SPEI from 1980 to 2018. The time lag correlation between meteorological drought events and the EI Nino-Southern Oscillation (ENSO), Arctic Oscillation (AO), North Atlantic Oscillation (NAO), and Pacific Decadal Oscillation (PDO), were analyzed in the LMRB. Our results indicated that, from a temporal perspective, the period from November to April of the following year was particularly prone to meteorological droughts in the basin. In terms of spatial distribution, the primary agricultural regions within the basin, including Thailand, Eastern Cambodia, and Vietnam, were highly susceptible to meteorological droughts. Further analysis revealed a teleconnection between drought events in the LMRB and atmospheric circulation factors. The sensitivity of the basin's drought timing to its response decreased in the order of the ENSO > AO > NAO > PDO. In general, the ENSO had the most substantial influence on drought events in the basin, with the strongest response relationship, while the upper reaches of the basin displayed the most significant response to the AO; the occurrence and progression of meteorological droughts in this area synchronized with the AO. These findings enhance our understanding of drought-prone areas in the LMRB, including the meteorological factors and driving mechanisms involved. This information is valuable for effectively mitigating and managing drought risks in the region.

**Keywords:** meteorological drought; atmospheric circulations; lag response; Lancang-Mekong River Basin

## 1. Introduction

A meteorological drought refers to an imbalance in the water balance caused by evaporation exceeding precipitation over a certain period, typically indicated by a shortage of precipitation [1]. The world is currently undergoing climate change, manifested by a rise in temperatures [2]. This temperature increase has led to a reinforcement of the water cycle, subsequently impacting precipitation, evaporation, and runoff [3]. Anomalous variations in ocean temperatures or atmospheric circulations are increasingly impacting regional climates [4]. This has amplified the frequency and severity of meteorological droughts, disrupting and posing threats to the economic development and ecological systems within the basin [5,6]. Therefore, it is imperative to delve into the spatio-temporal characteristics of meteorological droughts in the basin and unravel the intricate patterns governing their evolution [7]. These efforts are crucial for enhancing the basin's resilience against drought risks.

In terms of drought quantification, assessment, and monitoring, drought indices serve as crucial parameters for investigating drought characteristics [8]. Currently, the Palmer Drought Severity Index (PDSI) and the Standardized Precipitation Index (SPI) are widely employed as drought indices for monitoring and analyzing global and regional drought phenomena [9,10]. Nonetheless, in the backdrop of global change, the escalation in temperature has emerged as a significant factor intensifying the drought process. Consequently, an objective depiction of the drought situation necessitates the combined impact of alterations in precipitation and temperature [11]. The Standardized Precipitation Evapotranspiration Index (SPEI) provides an enhanced approach to evaluating drought conditions as it effectively incorporates the impacts of both precipitation patterns and potential evapotranspiration fluctuations. This comprehensive index seamlessly integrates the sensitivity of the PDSI to variations in evaporation demand, induced by temperature oscillations and trends. Additionally, it harnesses the simplicity of the computation associated with the SPI, while simultaneously leveraging its inherent multi-temporal and spatial characteristics [12]. This makes it an ideal instrument for monitoring desertification and studying the impact of warming on the desertification process [13,14]. The World Meteorological Organization (WMO) currently recognizes and endorses the SPEI as the primary meteorological indicator for monitoring meteorological droughts [14]. It is now widely employed across various global regions [13,15] and has demonstrated considerable applicability within the LMRB [16].

In terms of the mechanisms of drought formation, past research has primarily focused on issues such as the relative contribution of local land–atmosphere interactions to drought formation [17]. However, it is now recognized that atmospheric circulation factors, including the ENSO, NAO, AO, and PDO, can have a significant impact on drought formation over longer periods of time [18,19]. As a large-scale atmospheric phenomenon on a global scale, anomalies in atmospheric circulation can lead to sudden changes in temperature and evapotranspiration, altering the spatio-temporal distribution of precipitation and resulting in regional drought events. This climatic phenomenon is also known as teleconnection [20]. At the same time, atmospheric circulation has proven to have stable periodicity and persistence, and its impact on regional climate has a lag effect [21]. Therefore, the use of atmospheric circulation factors to predict meteorological drought events has become one of the hotspots, and has been used in China [19], Africa [18], Iran [22], and other countries [16], as well as in the Pearl Lake Basin [21], and the Yangtze River Basin [23]. Consequently, establishing a robust response relationship between atmospheric circulation patterns and meteorological drought indices within a basin can significantly contribute to a profound analysis of the mechanisms underlying drought occurrence and progression. This, in turn, facilitates effective intervention strategies and preventive measures against recurrent drought events in the basin [24].

The LMRB is globally recognized as one of the most crucial transboundary basins, offering invaluable water and energy resources, freshwater resources, and ecological resources to the countries within the basin [25,26]. It not only serves as the pillar of fishery development and food production in the riparian countries, but also acts as the economic link among the countries in the basin [27]. However, in recent years, the LMRB has been experiencing a worrisome escalation in both the frequency and severity of drought events, primarily attributed to the pervasive influence of global climate change. This distressing trend has resulted in far-reaching consequences for various sectors, including domestic water consumption, agricultural irrigation, ecological balance, and navigational activities across the countries encompassing the basin [28,29]. Thus, this study aimed to analyze the meteorological drought characteristics in the LMRB and identify the impact and correlation of atmospheric circulations on the meteorological drought in the basin. The specific research was as follows: (1) We analyzed the spatio-temporal variation characteristics of SPEI through the Run Theory and Mann-Kendall test to clarify the spatio-temporal evolution characteristics and changing trends of meteorological droughts in the LMRB. (2) By examining the time-lag correlation between atmospheric circulation indices and SPEI, we established a reliable basis for monitoring and predicting meteorological drought in the LMRB.

In comparison to previous research, the novelty of this manuscript is reflected in the following aspects. Firstly, this study diverges from the conventional approach of utilizing a 6-month time scale of SPEI. Instead, it takes into consideration the distinct wet and dry seasons in the LMRB, calculating the SPEI separately for the dry season (November to April) and the rainy season (May to October). This approach provides a more accurate depiction of the characteristics and trends of meteorological drought during both the dry and rainy seasons. Secondly, this study analyzed the teleconnections between atmospheric circulation patterns and drought and provided insights for predicting meteorological droughts in the LMRB. Furthermore, the research findings enable real-time monitoring of the synchronicity or lagged effects between the ENSO, AO, NAO, PDO, and meteorological drought. This allows for the implementation of effective measures in preventing and mitigating meteorological drought events within the basin. Ultimately, these findings serve as valuable scientific foundations for water resource management in the LMRB.

## 2. Materials and Methods

### 2.1. Research Area

The Lancang-Mekong River originates in Qinghai Province, China, traverses the Tibet Autonomous Region and Yunnan Province, and meanders through Myanmar, Laos, Thailand, Cambodia, and Vietnam, ultimately converging with the South China Sea to the west of Ho Chi Minh City, Figure 1a [30].

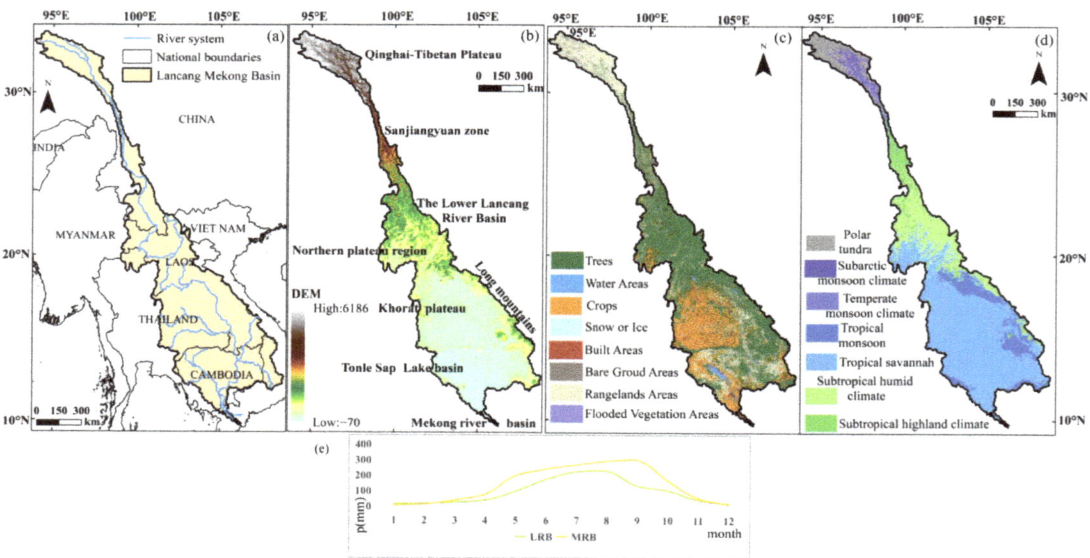


**Figure 1.** Location of the LMRB. (**a**) Location and water system distribution, (**b**) Elevation, (**c**) Land use types, (**d**) Climate type, (**e**) Average precipitation.

The LMRB covers an area of approximately 812,400 km$^2$, with a main stream that stretches about 4880 km in length, making it the first longest river in Southeast Asia [31]. Figure 1 illustrates the geographic location, water system distribution, elevation, and land use types. The rainy season (May to October) in the LMRB is dominated by the southwest monsoon from the ocean, resulting in a humid and rainy climate [32]. Conversely, during the dry season (November to April of the subsequent year), the basin is influenced by the northeast monsoon from the continent, leading to dry and less rainy conditions [7]. Generally, the annual precipitation upstream of the Lancang River is less than 1000 mm, while it is approximately 1500 mm downstream. The average annual precipitation in the Mekong River basin, on the other hand, exceeds 1500 mm [27]. The climate in the Lancang River basin exhibits a variety of climate types, with an annual average temperature range of 21 °C to 4.7 °C. In contrast, the temperature in the Mekong River basin shows a relatively uniform variation, with an annual average temperature ranging from around 25 °C to 27 °C [33]. The average annual runoff is approximately $4.75 \times 10^{11}$ $m^3$ of the LMRB, and the land utilization status in the LMRB exhibits regional variations, as depicted in Figure 1b,c. The Mekong Delta, Khorat Plateau, and Tonle Sap Lake Basin are the predominant agricultural cultivation areas in the region (Figure 1b,c) [34]. Owing to its unique geographical location, diverse climate types, and high sensitivity to climate change, the LMRB has become a focal point for global climate change research [27].

### 2.2. *Materials and Methods*

#### 2.2.1. Meteorological Data

This study utilized meteorological data, including precipitation and temperature, obtained from the Climatic Research Unit TS v.4.03 (CRU) database (https://crudata.uea.ac.uk/cru/data/hrg/cru_ts_4.03/ accessed on 1 January 2021). The CRU database provides monthly-scale data with a spatial resolution of 0.5° × 0.5°. The advantage of using the CRU dataset to calculate SPEI is that the CRU dataset is obtained based on a large number of site data, with excellent data quality, comprehensive coverage, and continuous time series [35]. Considering that calculating the SPEI index requires at least 30 years of meteorological data, the CRU database has sufficient data support in this regard [4].

The bias in the CRU data has been corrected using data from meteorological stations in the LMRB. Our approach entails interpolating CRU data to specific meteorological observation stations within the study area with data sourced from the China Meteorological Data Network. When choosing interpolation methods, we employed two prevalent techniques: inverse distance weighted interpolation and bilinear interpolation. The calculation methods or basic principles of the two interpolation methods selected in this study were as follows.

Specifically, the calculation method employed the inverse distance weighted interpolation as:

$$\hat{z} = \frac{\sum_{i=1}^{n} \frac{1}{(D_i)^P} Z_i}{\sum_{i=1}^{n} \frac{1}{(D_i)^P}} \tag{1}$$

In the equation above, $\hat{z}$ represents the estimated value, $Z_i$ represents the i (i = 1, …, n) sample, and $D_i$ represents the power of distance. P is a weighting factor that has a significant influence on the interpolation results. The selection criterion is based on the minimum average absolute error. Generally, higher powers result in a smoother interpolation effect. In this specific context, P was chosen as 2, which corresponded to the inverse squared distance.

The bilinear interpolation algorithm geometric operation is generally defined as follows:

$$g(x', y') = f(x, y) = f[a(x, y), b(x, y)] \tag{2}$$

In the equation above, $g(x', y')$ represents the output image, $f(x, y)$ represents the input image, and the function $a(x, y)$, $b(x, y)$ describes the spatial relationship between the two in the Cartesian coordinate system.

To evaluate the accuracy of these methods, we conducted an analysis based on statistical measures including the correlation coefficient I, mean absolute error (MAE), and root mean square error (RMSE). The specific calculation methods are as follows:

$$R = \frac{\sum_{i=1}^{T}(y_i - \overline{y}_i)^2}{\sqrt{\sum_{i=1}^{T}(y_i - \overline{y}_i)^2}} \quad (3)$$

$$MAE = \frac{|y_i - \overline{y}_i|}{T} \quad (4)$$

$$RMSE = \sqrt{\frac{1}{T}\sum_{i=1}^{T}(y_i - \overline{y}_i)^2} \quad (5)$$

In this context, "R" symbolizes the correlation coefficient, while "n" represents the sample size, and "$y_i$" and "$\overline{y}_i$", respectively, allude to the CRU data and the site data. The domain of "R" is [0,1], whereby values closer to 1 signify heightened data consistency. The range of MAE is $[0,-\infty]$, with an optimal value of 0. Similarly, the range of RMSE is $[0,-\infty]$, with an optimal value of 0.

These metrics were derived from the validation results and are presented in Table 1 to provide a comprehensive overview. The bilinear interpolation results were found to align more closely with the actual measurements (Table 1), leading us to select bilinear interpolation for calibrating anomalous values in this study.

**Table 1.** Verification of the bias correction results for the meteorological station data in the study area compared to the corresponding CRU precipitation and temperature data.

| Method | Data Type | R | MAE | RMSE |
|---|---|---|---|---|
| Distance weighted interpolation | Temperature | 0.756 | 8.21 | 10.56 |
| | Precipitation | 0.864 | 6.34 | 7.69 |
| The bilinear interpolation | Temperature | 0.951 | 2.32 | 4.82 |
| | Precipitation | 0.935 | 3.03 | 4.18 |

#### 2.2.2. Atmospheric Circulation Indices

Based on the current understanding of important teleconnections of atmospheric circulations to regional climate variability, this study selected the most commonly used four atmospheric circulation factors (Table 2). The monthly time series (1980–2018) for all these four atmospheric circulations were obtained from different sources and they are given in Table 2.

**Table 2.** Summary of atmospheric circulation indices used.

| NO | Atmospheric Circulation | ID | Data Source |
|---|---|---|---|
| 1 | EI Nino-Southern Oscillation | ENSO | NOAA Physical Sciences Division (PSD) |
| 2 | North Atlantic Oscillation | NAO | NOAA Climate Prediction Centre (CPC) |
| 3 | Arctic Oscillation | AO | NOAA Physical Sciences Division (PSD) |
| 4 | Pacific Decadal Oscillation | PDO | National Centre for Atmospheric Research (NCAR) |

#### 2.2.3. Standardized Precipitation Evapotranspiration Index

The LMRB exhibits well-defined seasonal variations, characterized by distinct dry and wet seasons. The wet season spans from May to October, while the dry season extends from November to the subsequent April. However, due to climate and environmental changes, meteorological droughts have become increasingly frequent in the region, affecting both the rainy and dry seasons and exacerbating the overall impact of drought [7]. The SPEI is suitable for monitoring the characteristics of meteorological droughts under global warming [36], due to the advantageous feature of SPEI encompassing multiple time scales [32]. Therefore, this paper calculated the SPEI on different time scales of 1 month, 6 months, and 12 months for 38 years, respectively. A one-month SPEI (SPEI-1) can indicate the occurrence of drought on a monthly basis. A three-month SPEI (SPEI-3) can depict the seasonal moisture conditions. A six-month SPEI for both the wet and dry seasons can, respectively, represent the moisture conditions of the basin during these periods. A twelve-month SPEI (SPEI-12) can demonstrate the persistence of interannual moisture conditions.

The SPEI index calculation method in this paper is based on Vicente-Serrano [37]. Given that the Thornthwaite method is more practical in the absence of complete and high-quality meteorological data [38], this paper utilized the Thornthwaite method to compute the potential transpiration (PET) for the LMRB.

Subsequently, the disparity between monthly precipitation and potential evapotranspiration was computed:

$$D_m = P_m - ET_{0m} \tag{6}$$

where, m is the number of months, $P_m$ is the monthly precipitation, and $ET_{0m}$ is the potential evapotranspiration.

Third, $D_m$ is aggregated and normalized according to different time scales:

$$\begin{cases} D^i_{m,n} = \sum_{j=13-i+n}^{12} D_{m-1} + \sum_{j=1}^{n} D_{m,j} & ,n < i \\ D^i_{m,n} = \sum_{j=n-i+1}^{n} D_{m,j} & ,n \geq i \end{cases} \tag{7}$$

Then, fitting $D_m$ using log-logistics:

$$F(D) = \left[1 + \left(\frac{\alpha}{D-\gamma}\right)^{\beta}\right]^{-1} \tag{8}$$

where, the parameter $\alpha$, $\beta$, and $\gamma$ denote the scale, shape, and position parameters estimated by linearity, respectively.

Finally, we can normalize the cumulative probability density:

$$\begin{aligned} SPEI &= W - \tfrac{c_1+c_2w+c_3w}{1+t_1w+t_2w^2+c_3w^3} \\ W &= \sqrt{-2\ln(p)} \end{aligned} \tag{9}$$

In the equation, when $|p| \leq 0.5$, $p = 1 - p$; and when $p > 0.5$, $p = 1 - p$, $c_1 = 2.515517$, $c_2 = 0.802853$, $c_3 = 0.010328$, $t_1 = 1.432\,788$, $t_2 = 0.189269$, $t_3 = 0.001308$ [10].

Considering the geographical attributes of the LMRB along with relevant research, the drought conditions prevailing in this region are categorized into five distinct levels based on the SPEI, as presented in Table 3.

**Table 3.** The classification of meteorological drought and wetness grade based on SPEI.

| Level | Type | SPEI Value |
|---|---|---|
| 1 | Normal/Wetness | $-0.5 <$ SPEI |
| 2 | Mild drought | $-1.0 <$ SPEI $\leq -0.5$ |
| 3 | Moderate drought | $-1.5 <$ SPEI $\leq -1.0$ |
| 4 | Severe drought | $-2.0 <$ SPEI $\leq -1.5$ |
| 5 | Extreme drought | SPEI $\leq -2.0$ |

### 2.2.4. The Run Theory

The Run Theory, initially introduced by Yevjevich [39], has gained widespread acceptance in the identification and characterization of drought events. In this study, a drought event is defined based on three predetermined criteria: (1) a persistent SPEI < 0, (2) a duration of at least 2 months during the period of negative SPEI values, and (3) a minimum

SPEI < −1 [40]. According to the run theory, the severity of a drought is determined by summing the SPEI values that fall below the specified threshold level [41,42].

Drought duration refers to the consecutive period during which the SPEI value remains below the specified threshold level. The frequency of drought is determined by counting the occurrences when the SPEI falls below the threshold level. The Run theory is a widely recognized approach for identifying drought events. In this study, particular attention was given to three essential factors for drought risk analysis: drought duration (Dd), drought severity (Ds), and drought peak (Dp) [7]. Figure 2 depicts the three meteorological drought events and their key features, as identified through the application of the Run Theory.

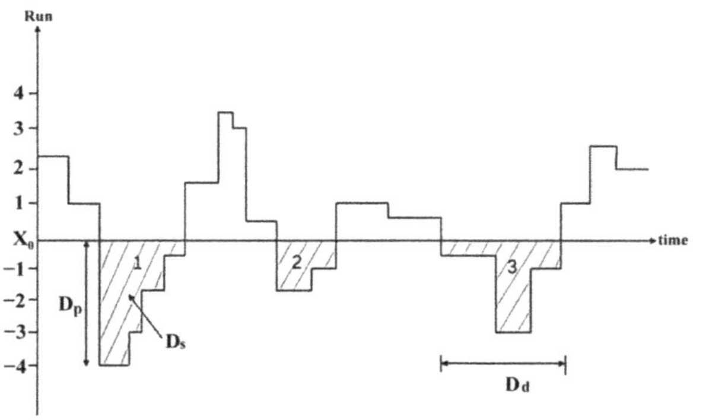


**Figure 2.** Run Theory identifies meteorological drought variables (1, 2, 3 respectively represent three distinct meteorological drought events identified using the Run Theory).

2.2.5. Mann-Kendall Trend Test

The non-parametric Mann–Kendall (M–K) test, initially introduced by Mann and Kendall, is widely endorsed and recommended by the WMO [43]. This test is valuable for evaluating trends or changes in meteorological time series, as it does not rely on specific distribution assumptions for the samples [16]. This study employs the M–K trend testing method to compute and analyze the trend of the SPEI in the LMRB, calculated as follows [44]:

$$Q = \sum_{j=i}^{n-1} \sum_{k=j+1}^{n} \mathrm{sgn}(x_k - x_j) \tag{10}$$

$$\mathrm{Var}(Q) = \frac{n(n-1)(2n+5) - \sum_{i=1}^{p} t_i(t_i-1)(2t_i+5)}{18} \tag{11}$$

$$Z = \begin{cases} \frac{Q-1}{\sqrt{\mathrm{Var}(T)}} & Q > 0 \\ 0 & Q = 0 \\ \frac{Q+1}{\sqrt{\mathrm{Var}(Q)}} & Q < 0 \end{cases} \tag{12}$$

The test statistic used in this study is denoted as 'Q', representing the number of tied groups in the data, which indicates the frequency of data repetition. The length of the data series is represented by 'n', and 'x, j' refers to the data values at times 'i' and 'j'. The standardized statistic for the Mann–Kendall test is denoted as 'Z', with positive and negative values indicating upward and downward trends, respectively. The trend is deemed statistically significant at the 0.05 (or 0.01) significance level when the absolute value of the Z-score, denoted as $|Z|$, exceeds or equals 1.96 (or 2.33) respectively [36].

2.2.6. Frequency of Meteorological Drought

In this study, we first identified drought events based on the Run Theory for the time scales of rainy season, dry season, and interannually using the SPEI. Secondly, using the results obtained, we calculated the occurrence frequencies of drought events for different severity levels according to Equation (13).

$$P_i = \left(\frac{n_i}{N}\right) \times 100\% \quad (13)$$

In the above equation, "$P_i$" is used to evaluate the frequency of drought occurrence at a specific grid point in the study area from 1980 to 2018. "N" represents the total number of years considered which is N = 39. "$n_i$" represents the number of years when drought occurred at the "i-th" grid point. The occurrence frequencies for each drought severity level are calculated based on the number of years of occurrence for each level.

2.2.7. Cross-Correlation Function

The Cross-correlation Function (CCF) is a valuable tool for assessing the resemblance between two variables at various time lags [45]. This method enables the determination of correlation coefficients between variables and previous occurrences [10]. One notable advantage of the cross-correlation approach is its ability to assess correlation coefficients at various lags, including positive and negative lags [46]. In this study, the CCF coefficient was used to examine asynchronous relationships between variables, considering lag times from 0 to 12 months. The significance of these correlations was determined using a *t*-test at a confidence level of 95%. Specifically, CCF was utilized to examine the lagged impact of ocean-atmospheric oscillation patterns on drought indices. The CCF can be formulated as follows [47]:

$$r_a = \frac{\check{C}_a(x,y)}{\check{\delta}_x \check{\delta}_{y+a}} \quad (14)$$

whereas the covariance and standard deviation of the sample are expressed as follows, respectively:

$$\begin{cases} \check{C}_a = \frac{1}{n-a}\sum\limits_{i=1}^{n-a}(x_i-\overline{x})(y_{i+a}-\overline{y}_{i+a}) \\ \check{\delta}_x = \left[\frac{1}{n-k}\sum\limits_{i=1}^{n-a}(x_i-\overline{x}_i)^2\right]^{\frac{1}{2}} \\ \check{\delta}_{y+a} = \left[\frac{1}{n-a}\sum\limits_{i=1}^{n-a}(y_{i+a}-\overline{x}_{i+a})^2\right]^{\frac{1}{2}} \end{cases} \quad (15)$$

Additionally, the mean value is:

$$\begin{cases} x_i = \frac{1}{n-a}\sum\limits_{i=1}^{n-a} x_i \\ \overline{y}_{i+a} = \frac{1}{n-a}\sum\limits_{i=1}^{n-a} \overline{y}_{i+a} \end{cases} \quad (16)$$

where, $x_i$ is the number of sample sequences and $\overline{y}_i$ denotes the time lag with a unit of month. As a rule of thumb, the absolute value of the time delay 'a' should be less than n/4. It is noteworthy that this paper employs Python to conduct a 0–12 month lag cross-correlation analysis on the SPEI-1 of each grid point within the study area from 1980 to 2018 and the monthly atmospheric circulation data. Owing to the extensive data set, the range of correlation coefficient values was somewhat reduced. Studies indicate that although the correlation coefficients are low, their ability to pass the significance test means that we can discount results that are due to chance [24,44]. Thus, affirming the meaningfulness of the relationship between drought events and atmospheric circulation factors.

## 3. Results

### *3.1. Temporal Variation Characteristics of LMRB Drought*

This study identified meteorological drought events and their characteristics in the LMRB using the Run Theory. Figure 3 illustrates the variations in SPEI at 1, 3, 6, and 12 months throughout the study period (1980–2018).

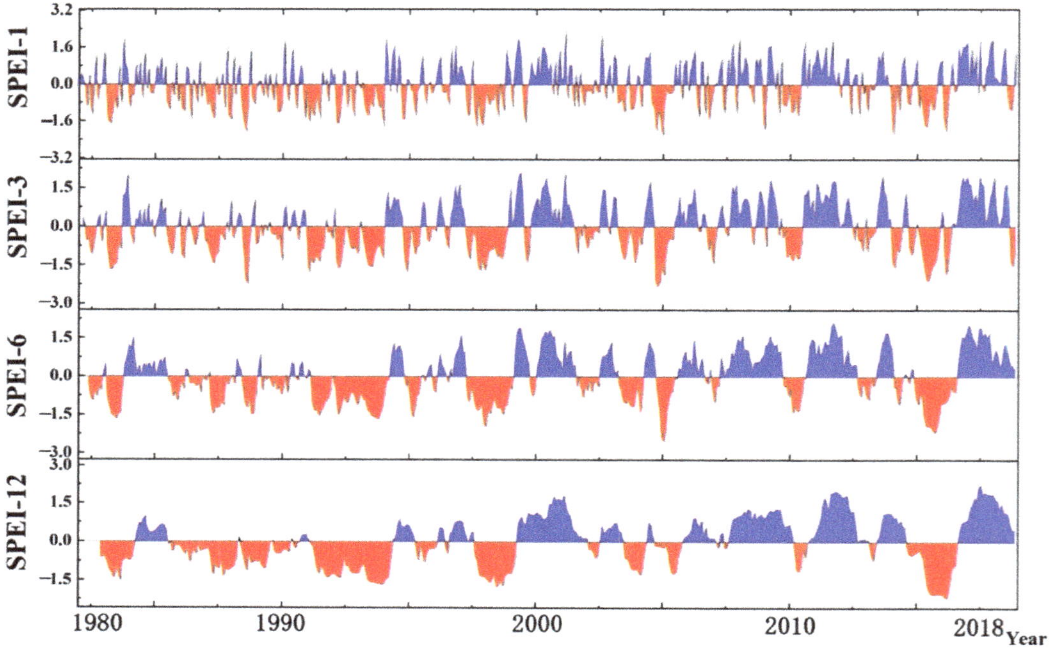


**Figure 3.** Temporal variation of SPEI at different time scales in the LMRB-based Run Theory (Blue represents positive values of SPEI, while red represents negative values of SPEI).

As illustrated in the figure, the SPEI showcased distinct oscillations between positive and negative values, particularly on the temporal scales of 1 month and 3 months. Generally, with an increase in the time scales, the cumulative effects of the SPEI became more obvious with longer wet and dry periods. This observation implies that the LMRB undergoes frequent transitions between periods of dryness and wetness, both on a monthly and seasonal timeframe. From November to April of the subsequent year, the SPEI consistently showed negative values, indicating a period that was susceptible to meteorological drought. Moreover, it is evident that since 1990, there is a potential for meteorological drought events to occur in all months of the year, and the risk of meteorological drought in summer and autumn has escalated. From SPEI-6 and -12, it is apparent that the basin had significant and continuous negative values in 1983, 1987–1989, 1991–1993, 1998, 2004, 2005, 2009–2010, 2012–2013, and 2015–2016. This suggests that meteorological drought events may have transpired in these years. Furthermore, we have statistically analyzed the duration, severity, and intensity of drought across different SPEI time scales, culminating in the results presented in Table 4. Our findings indicate that with the extension of the SPEI time scale, both the duration and severity of meteorological drought events in the LMRB tend to increase. The average durations of SPEI-1, SPEI-3, SPEI-6, and SPEI-12 are 6, 10, 13, and 18 months, respectively. By classifying meteorological drought levels based on the SPEI (as shown in Table 3), and considering the temporal variability of SPEI-6 and SPEI-12 as depicted in Figure 3, it was revealed that for the meteorological drought events occurring in the LMRB during the periods of 1991–1993, 1998, and 2015–2016 (Table 4), the severity of these drought events surpassed the average level of previous years.

**Table 4.** Historical drought characteristics of SPEI at 1-, 3-, 6- and 12-month time scales.

| Time Scale | Number of Events | Average Duration | Average Severity | Average Intensity |
|---|---|---|---|---|
| 1 | 82 | 6 | 4.57 | 1.33 |
| 3 | 50 | 10 | 8.06 | 1.43 |
| 6 | 29 | 13 | 10.45 | 1.67 |
| 12 | 17 | 18 | 13.56 | 1.45 |

### 3.2. Spatial Variation Trend of LMRB Drought

In light of the distinct wet and dry seasons characteristic of the LMRB, this study proceeded to calculate the SPEI for both seasons within the basin, with a temporal resolution of May to October (wet season) and November to April of the following year (dry season), each spanning a 6-month period. To differentiate from the previously mentioned SPEI-6, the SPEI values for the dry and wet seasons in this study are denoted as SPEI-dry and SPEI-wet, respectively. Based on the drought classification criteria, this study categorized the drought events in the LMRB between 1980 and 2018 into various levels of drought severity. In particular, a drought event is categorized as mild when the SPEI value drops below $-0.5$ but remains above $-1.0$. A moderate drought is identified when the SPEI value is less than $-1.0$ but more than $-1.5$. A severe drought is recognized when the SPEI value is less than $-1.5$ but more than $-2.0$, and an extreme drought event is characterized by an SPEI value below $-2.0$. This section calculates and analyzes the spatial variations of SPEI-dry, SPEI-wet, and SPPEI-12 during the period of 1980–2018. The objective of this research was to portray the frequency of drought occurrences in the LMRB from 1980 to 2018 by evaluating the proportion of drought events across different levels of severity during the specified study period. Figure 4 visually demonstrates the spatial distribution of drought frequency at various severity levels within the LMRB.

By analyzing the occurrence frequencies of different levels of meteorological droughts during different time periods (rainy season, dry season, and interannual), it can be observed that the frequency of mild meteorological drought was the highest in this basin, followed by moderate, severe, and extreme meteorological drought events. Furthermore, the dry season served as the peak period for meteorological drought events. The frequencies of mild, moderate, severe, and extreme meteorological drought occurrences were 17.94% to 51.28%, 0% to 25.64%, 0% to 20.51%, and 0% to 10.25%, respectively (Figure 4e–h).

It is worth noting that the occurrence frequency of meteorological droughts was higher during the rainy season. The frequencies of mild, moderate, severe, and extreme meteorological drought occurrences during this period were 12.84% to 41.08%, 0% to 23.07%, 0% to 17.94%, and 0% to 10.25%, respectively, as shown in Figure 4a–d. Further statistical analysis and comparison of the different levels of meteorological drought occurring during the rainy season and dry season within the basin revealed the following patterns: After further statistical comparison of different levels of meteorological drought occurring during the wet and dry seasons within the basin, we found that during the wet season, the Three Rivers Source Region, the lower of the Lancang River, the Northern Plateau, and the Tonle Sap Lake Basin were high-incidence areas for mild meteorological drought; the Khorat Plateau was a high-incidence area for severe meteorological drought events, with a low frequency of extreme drought during the wet season, and occasional occurrences of extreme meteorological drought in the Long Mountains area, as shown in Figure 4a–d.

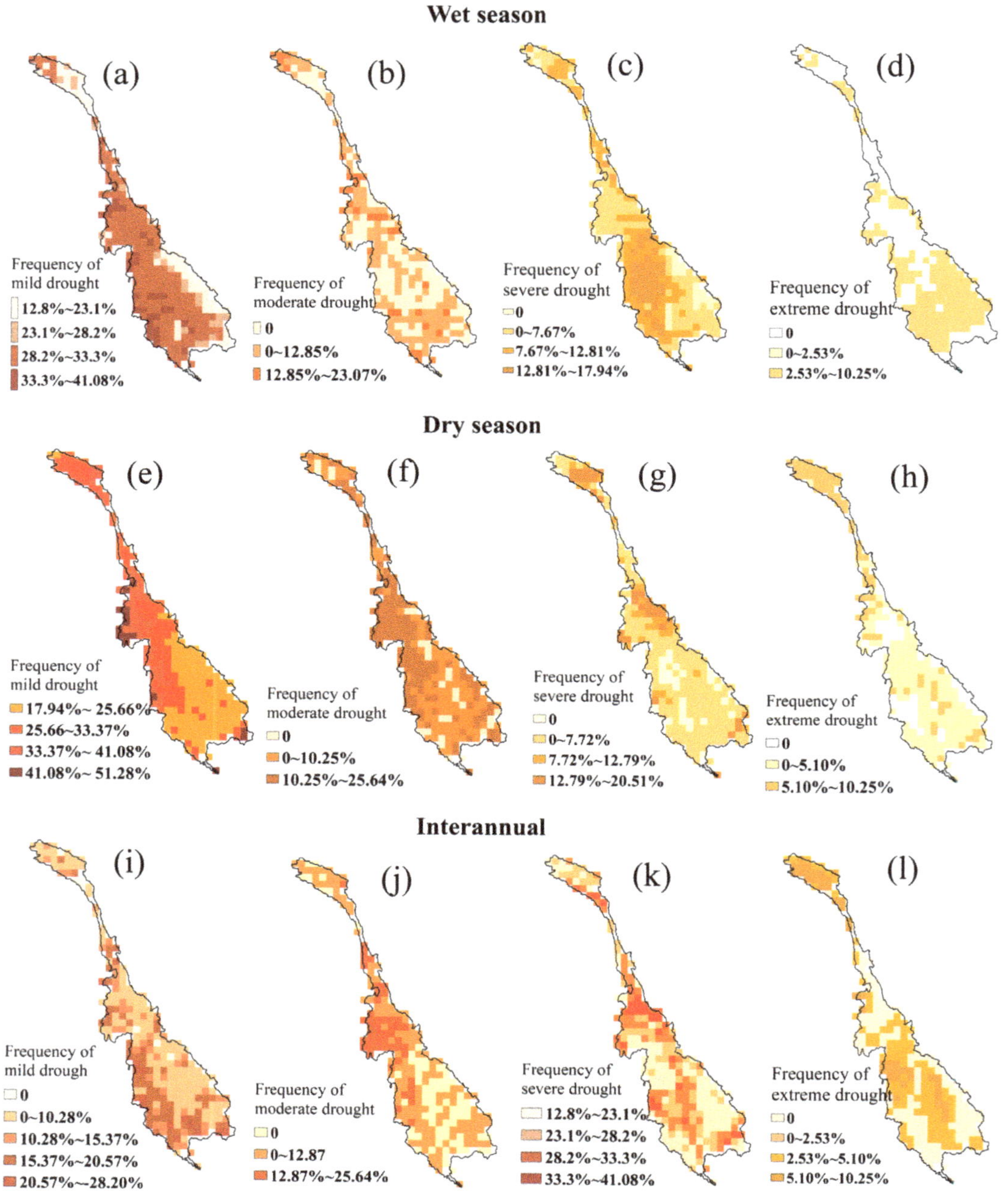


**Figure 4.** The spatial variation of meteorological drought frequency at different severity levels in the Lancang-Mekong River Basin ((**a**–**d**): variation of meteorological drought frequency during the wet season, (**e**–**h**): variation of meteorological drought frequency during the dry season, (**i**–**l**): variation of meteorological drought frequency in interannual).

However, the spatial differentiation of different levels of meteorological drought events was more pronounced during the dry season. As can be seen from Figure 4e–h, the western parts of the Lancang River Basin, the Northern Plateau, and the Khorat Plateau were high-incidence areas for mild meteorological drought (Figure 4e) and severe meteorological drought (Figure 4f); the high-incidence area for severe meteorological drought events was the eastern part of the Northern Plateau (Figure 4g). Compared to the rainy season and dry season, at the annual scale within the basin, the occurrence frequency of different levels of meteorological drought was relatively average. The high-frequency zone for mild meteorological drought was located in the western parts of the Khorat Plateau and the Tonle Sap Lake Basin, as shown in Figure 4i. The western part of the Northern Plateau was the high-frequency zone for moderate meteorological drought (Figure 4j), while the lower reaches of the Lancang River were the high-frequency zone for severe meteorological drought (Figure 4k). By comparing the occurrence frequencies of meteorological drought in the same region at different time scales, we obtained the following information. Whether during the rainy season or dry season, the occurrence frequency of mild meteorological drought in the Lancang River basin was higher than in other areas of the basin, with an average occurrence rate of approximately 25%. The occurrence frequencies of moderate and severe meteorological droughts in the downstream area of the basin were higher than in the upstream area, primarily concentrated in the western parts of the Northern Plateau, Khorat Plateau, and Tonle Sap Lake Basin, as shown in Figure 4c,f,g.

It is worth noting that the upper reaches of the Lancang River basin, Khorat Plateau, Mekong Delta, and the central part of the Tonle Sap Lake Basin were high-frequency zones for extreme meteorological drought events. We observed that extreme meteorological drought events primarily occurred in the Khorat Plateau, Tonle Sap Lake Basin, and Mekong Delta during the rainy season, with occurrence frequencies that ranged from 2.5% to 10.25%, as shown in Figure 4d. On an annual scale, the upper reaches of the Lancang River Basin, and the northern and eastern parts of the Khorat Plateau, had a frequency of extreme meteorological drought events ranging from 2.5% to 10.25%, which was higher than other areas, as shown in Figure 4i. In summary, the LMRB has the highest frequency of mild meteorological drought, with nearly the entire basin experiencing such events from 1980 to 2018. At the same time, the high-incidence areas for moderate and severe meteorological drought were concentrated in the Northern Plateau and the Khorat Plateau. Extreme meteorological drought events (SPEI $< -2$) occurred throughout the basin between 1980 and 2018, with a frequency of 0 to 10.26%, and the incidence of extreme drought events (SPEI $< -2$) was notably prevalent in the upper reaches of the Lancang River, Khorat plateau, Tonle Sap Lake Basin, and the Mekong Delta (Figure 4c,f,j,k). It is noteworthy that despite the LMRB being situated in the heart of the Asian tropical monsoon region, and being influenced by the southwest monsoon from May to the end of September characterized by high humidity and heavy rainfall, meteorological drought events still persisted (Figure 4a–h).

### 3.3. Spatial Variation of Drought Trends in Dry and Wet Season

Based on the aforementioned analysis, it is evident that meteorological drought occurrences are apparent in the LMRB, even during the rainy season. Therefore, we calculated and analyzed the trends of SPEI-dry, SPEI-wet, and SPEI-12 using the Mann–Kendall trend test. This study further analyzed the dry-wet trend in the basin during the dry season and rainy season, and annually (Figure 5).

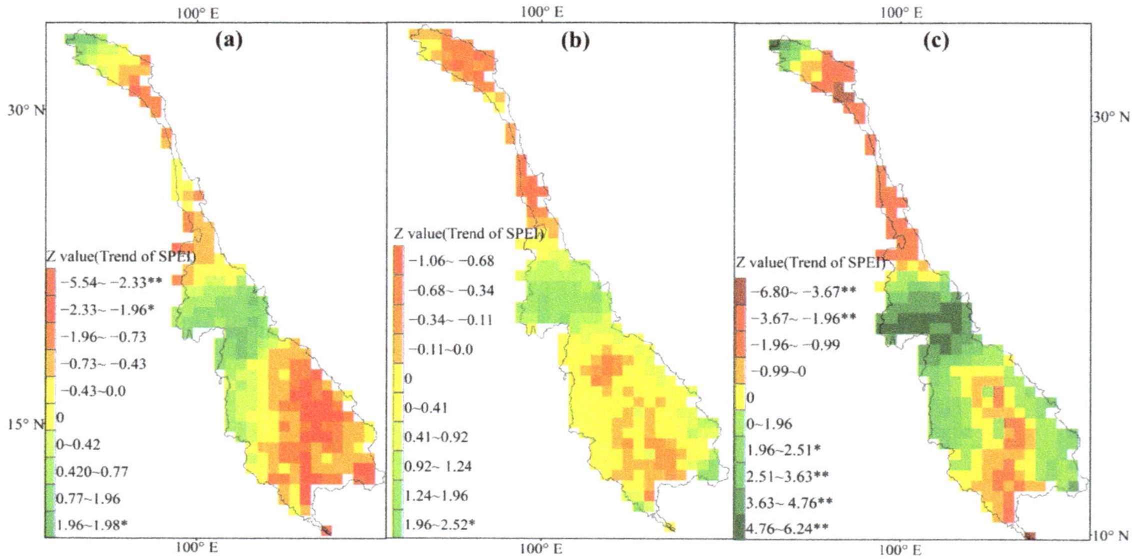


**Figure 5.** Spatial variation of meteorological drought and wetness trends in the LMRB (**a**) wet season (**b**) dry season (**c**) annual, * represents that the trend of SPEI has passed the significance test at the 0.05 level. ** represents that the trend of the SPEI has passed the significance test at the 0.01 level.

From the preceding analysis, it was evident that meteorological drought events transpired in the LMRB even during the rainy season. Therefore, this study further analyzed the dry-wet changes in the basin during the dry season and rainy season, and annually (Figure 5a,b). During the rainy season, the drought trends in the Long Mountains region of the basin, as well as in the Tonle Sap Lake Basin and Mekong Delta, were statistically significant ($p < 0.01$), while the Northern Plateau region showed a significant moistening trend ($p < 0.05$), as shown in Figure 5a. During the dry season, the upper reaches of the LMRB, as well as the Khorat Plateau and the central part of the Tonle Sap Lake Basin, exhibited a drying trend, while the Northern Plateau showed a moistening trend, although not statistically significant, as shown in Figure 5b. This also indicated that the changes in wetness during the rainy season were more pronounced in the LMRB, as shown in Figure 5a. Further investigation showed that, regardless of the season, the northern plains within the basin (encompassing Myanmar, southern Laos, and southern Thailand) demonstrated a significant trend towards wetter conditions, as shown in Figure 5a,b. On an annual scale, the dry-wet changes in the LMRB aligned closely with those observed during the wet season, as shown in Figure 5c. The middle and lower reaches of the Lancang River, the eastern segment of the Khorat Plateau, the Tonle Sap Lake Basin, and the Mekong Delta revealed a discernible propensity towards drought ($p < 0.01$), while the northern plateau exhibited a pronounced trend towards wetter conditions ($p < 0.01$). The dry-wet change trends in other areas were not significant, as shown in Figure 5c. Therefore, overall, the changes in dryness and wetness in the basin showed spatial heterogeneity. The northern region has abundant precipitation and large river flow, and the trend of wet changes is becoming more significant. Nevertheless, during both the dry and rainy seasons, key areas with elevated water demand, such as fisheries and agriculture, exhibit a noteworthy trend towards desertification, such as the Mekong Delta. Therefore, in these regions where there is a significant differentiation between dryness and wetness, the possibility of suffering from drought/flood disasters will increase.

*3.4. Response of the Meteorological Drought to Atmospheric Circulations*

The influence of atmospheric circulation factors on meteorological elements such as precipitation and temperature is a continuous process. It requires a certain amount of time to culminate in a meteorological drought event, a phenomenon referred to as "lag response" [24]. Therefore, this study aimed to explore the long-distance connection between the changes in monthly atmospheric circulation factors (ENSO, NAO, AO, and PDO) and meteorological drought events (SPEI-1) in the LMRB. This was achieved by establishing the lag correlation between these factors and meteorological drought events and visualizing the spatial distribution using GIS.

As depicted in Figures 6–9 the lag correlation between atmospheric circulation factors and meteorological drought events in the LMRB demonstrated a descending order of response strength from the ENSO, AO, and NAO to PDO. Specifically, the meteorological drought events in the LMRB were most sensitive to the ENSO, as shown in Figure 6. In particular, the western regions of the Mekong River basin (western Northern Plateau, western Khorat Plateau, western Tonle Sap Lake Basin, and Mekong Delta) exhibited synchronous meteorological drought events with the ENSO, showing a negative correlation ($p < 0.05$), as shown in Figure 6 (lag-0). From the spatial distribution of the lagged cross-correlation between the ENSO and meteorological drought, we observed that meteorological drought in the Lancang River basin exhibited a positive lagged correlation with the ENSO. On the other hand, meteorological drought in the Mekong River basin showed a negative lagged correlation with the ENSO.

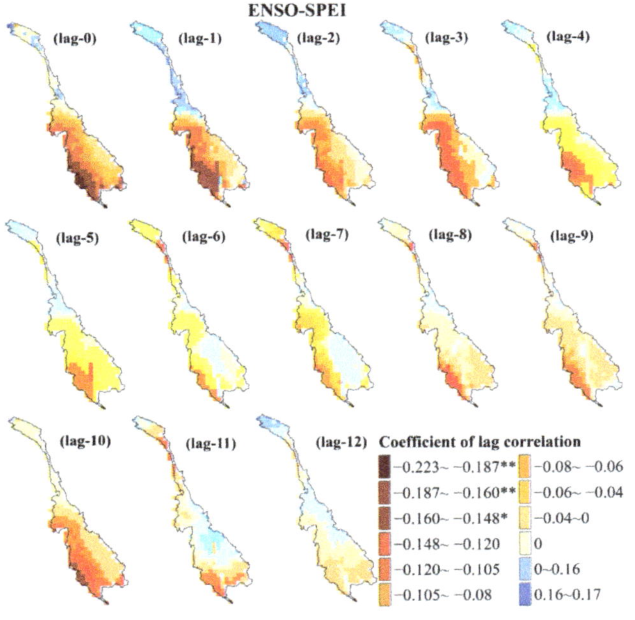


**Figure 6.** Distribution of correlation coefficients between the meteorological drought and NAO in the LMRB (Lag-0, lag-1, lag-3, . . . lag12 represents lagged 0, 1, 2, 3 . . . 12 months between meteorological drought and ENSO, * represents that correlation coefficient has passed the significance test at the 0.05 level, ** represents that correlation coefficient has passed the significance test at the 0.01 level).

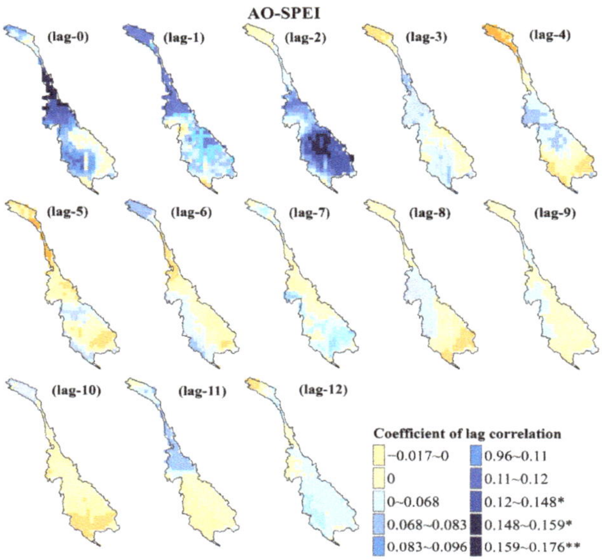


**Figure 7.** Distribution of correlation coefficients between the meteorological drought and AO in the LMRB (Lag-0, lag-1, lag-3, ... lag12 represents lagged 0, 1, 2, 3 ... 12 months between meteorological drought and AO, * represents that correlation coefficient has passed the significance test at the 0.05 level, ** represents that correlation coefficient has passed the significance test at the 0.01 level).

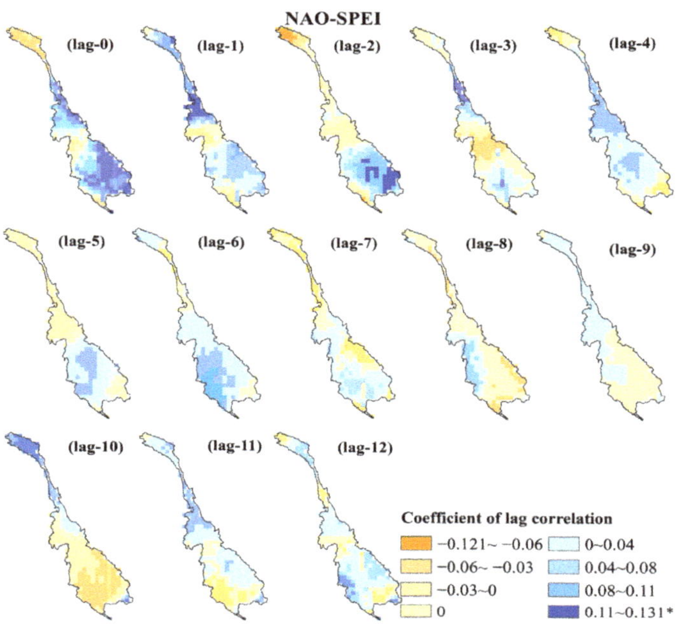


**Figure 8.** Distribution of correlation coefficients between the meteorological drought and NAO in the LMRB (Lag-0, lag-1, lag-3, ... lag12 represents lagged 0, 1, 2, 3 ... 12 months between meteorological drought and NAO, * represents that correlation coefficient has passed the significance test at the 0.05 level).

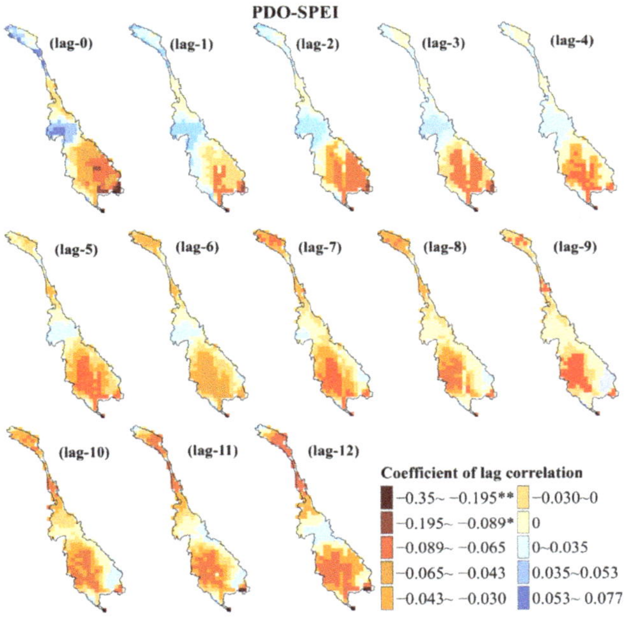


**Figure 9.** Distribution of correlation coefficients between the meteorological drought and PDO in the LMRB (Lag-0, lag-1, lag-3, . . . lag12 represents lagged 0, 1, 2, 3 . . . 12 months between meteorological drought and PDO, * represents that correlation coefficient has passed the significance test at the 0.05 level, ** represents that correlation coefficient has passed the significance test at the 0.01 level).

Specifically, the strongest negative lagged correlation was observed on the western bank of the Mekong River, and the lagged correlation gradually decreased from west to east along the western bank of the Mekong River, as shown in Figure 6 (lag-0–lag-5). As the lag time between the ENSO and meteorological drought increased, the area exhibiting a positive lagged correlation between the two gradually expanded. However, the significance level did not pass the statistical test, as shown in Figure 6 (lag-0–lag-12).

The lagged positive correlation between the AO, NAO, and the meteorological drought events in the LMRB was more significant, as shown in Figures 7 and 8. The meteorological drought in the LMRB exhibited the strongest positive correlation with the AO at a lag of 0–2 months, as depicted in Figure 7 (lag-0–lag-2). As the lag time between meteorological drought and the AO increased, the lagged negative correlation between the two decreased, as shown in Figure 7 (lag-3–lag-12). In the lower regions of the Lancang River basin, meteorological drought events exhibited synchronous behavior with the AO at a significant level ($p < 0.01$), as illustrated in Figure 7 (lag-0). Conversely, in the Khorat Plateau, Long Mountains, and Tonle Sap Lake Basin of the Mekong River basin, meteorological drought events displayed a significant correlation with a lag of 2 months to the AO ($p < 0.01$), as depicted in Figure 7 (lag-2).

In contrast to the AO, although meteorological drought in the LMRB exhibited a relatively strong positive correlation with the NAO at a lag of 0–2 months, the area demonstrating significantly lagged cross-correlation was smaller, as shown in Figure 7 (lag-0–lag-2) and Figure 8 (lag-0–lag-2). Notably, a significant positive lagged correlation was observed between meteorological drought occurrences in the lower regions of the Lancang River basin and the southern Long Mountains with the NAO, with respective lag times of 1 and 2 months, as depicted in Figure 8 (lag-1–lag-2). As the lag time increased, the area within the basin exhibiting a positive correlation between meteorological drought and the NAO gradually expanded. When the lag time reached 10 months, the region with a positive correlation between the two variables was maximized, although it did not pass the significance test, Figure 8 (lag-0–lag-10). The lagged negative correlation between the PDO and the meteorological drought events in the LMRB was more significant. In meteorological drought events in the lower reaches of the Mekong River, particularly in the northern part of the Long Mountains, there was synchrony with the PAO, displaying a significant negative correlation ($p < 0.01$), as illustrated in Figure 9. Based on the research results, monitoring the ENSO, AO, NAO, and PDO, and their synchronization or lag with meteorological drought can help in proactively intervening and preventing drought events in the basin.

## 4. Discussion

This study utilized CRU data and applied the multi-time scale SPEI (1-, 3-, 6-, 12-month) and run theory to analyze the spatio-temporal variation characteristics of meteorological drought in the LMRB. The findings revealed that meteorological drought events occurred frequently in the basin in recent years, particularly during the periods of 2004–2005, 2009–2010, 2012–2013, and 2015–2016, with long durations and severe drought conditions. These results are consistent with the historical natural disaster records of the LMRB in the international disaster database (https://public.emdat.be/data/ accessed on 1 October 2023) and with the results of previous studies [27,29]. In addition, by comparing the frequency and spatial distribution characteristics of meteorological drought events in the rainy season and dry season, and interannually in the LMRB, it was observed that even during the rainy season, meteorological drought events still occurred in the basin. During the dry season, the frequency of meteorological droughts increased, particularly impacting the upstream areas of the basin, such as Thailand, eastern Cambodia, and Vietnam. These findings are consistent with previous results [31,48], and they also verify the applicability of the research methods and data used in this study in the LMRB.

The LMRB is a crucial transboundary river, and the socio-economic development of the riparian countries is heavily reliant on its water resources, with agricultural usage comprising over 80% of the total water consumption [4,42]. However, due to uneven water resource distribution and differences in technological levels, the overall development level of water resources in the basin is not high [28,49]. Meanwhile, the uneven distribution of precipitation in the LMRB, compounded by the impact of climate change, has led to frequent meteorological drought events in the basin. These events pose severe threats and have significant impacts on socio-economic development and production activities within the basin [25,30]. According to the land use situation in the LMRB (Figure 1c), and the spatial characteristics of the frequency of meteorological drought events of different severities (Figure 4), it can be seen that the Khorat Plateau and the Tonle Sap Lake Basin, as part of the Mekong Delta region, are agricultural planting areas. However, these areas frequently experience meteorological drought events, especially the Mekong Delta region, which shows a significant trend of aridification (Figure 5), increasing the risk of agricultural planting.

According to the irrigation water use in the dry and rainy seasons in 2018 (Table 5), it can be seen that in 2018, the irrigation water used in the dry season accounted for the majority of the total water use in all basin countries except Thailand. However, the water production within each basin country can support the current irrigation demand for crop development so far (Table 5). Although Laos has abundant surface water resources, there are few agricultural irrigation projects. Currently, there are over 700,000 hectares of arable land, with a maximum irrigation area of 150,000 hectares during the wet season and only 4% of the arable land can be irrigated during the dry season. Thailand has an existing irrigation area of about 500,000 hectares, with an actual irrigation rate of less than 6% for arable land. Vietnam's Mekong River Basin, within its territory, has rich water energy reserves. Due to the serious problem of seawater intrusion in Vietnam, a large amount of freshwater resources need to be consumed, leaving a small amount of water available for use. Therefore, Vietnam mainly utilizes the water resources of the Mekong

River for agricultural irrigation, with about 2.4 million hectares of arable land and only 500,000 hectares of irrigable area. These are issues that deserve attention, especially during meteorological droughts, which can pose greater risks. Vietnam's water use in the dry season has already exceeded its capacity. Considering that the dry season is the peak period for agricultural water use in the Mekong River Basin [25,32], and our research has found that the main areas with high water demand for fisheries and agriculture in the basin in recent years have shown a significant trend of drought, these areas are more likely to be affected and threatened when a meteorological drought occurs in the basin. Therefore, it is recommended that all regions in the basin properly manage the seasonal allocation of water resources to ensure the agricultural production and water resource security of each community and improve their ability to cope with Meteorological drought [50,51].

**Table 5.** The water resource and irrigation water of countries along the LMRB in 2018.

| Type | Time | China | Myanmar | Laos | Thailand | Cambodia | Vietnam |
|---|---|---|---|---|---|---|---|
| Basin area ($10^4$ km$^2$) | / | 16.5 | 2.4 | 20.2 | 18.4 | 15.5 | 6.5 |
| average water yield (m$^3$/s) | / | 2410 | 300 | 5270 | 2560 | 2860 | 1660 |
| Irrigated area ($10^4$ hm$^2$) | / | 42 | — | 70 | 50 | 58 | 240 |
| Irrigation water consumption ($10^8$ m$^3$) | Dry season | 20.48 | — | 34.4 | 63.23 | 79.59 | 193.75 |
| | Rainy season | 5.12 | — | 11.47 | 94.84 | 34.11 | 68.07 |

/: nothing —: no data

The prediction of meteorological drought occurrence and development is of great significance for implementing drought mitigation measures and ensuring the sustainable development of water resources and ecosystems in the LMRB [33]. The ENSO phenomenon is the most important predictable source of drought in many areas worldwide, characterized by its stable cycle and strong persistence [52]. Our results also indicate that the meteorological drought in the LMRB is most sensitive to the ENSO, showing synchronicity. Therefore, the ENSO can serve as a crucial early warning signal for meteorological drought in the basin, effectively forecasting and intervening in meteorological drought, thereby providing support for drought assessment, drought prediction, and early warning in this region [53,54]. It is recommended to fully utilize the synchronicity between meteorological drought in the LMRB and the ENSO for effective monitoring and forecasting of meteorological drought [55]. Long-term drought forecasts are crucial for proactive measures, including adjusting and optimizing crop planting strategies. For regions at high drought risk, planting drought-resistant crops is recommended [56]. In areas with medium to low drought risk, it is advisable to consider planting crops with higher water demand [42]. Additionally, it is important to expedite the selection and cultivation of drought-resistant crops with specific genes [57,58]. Modern technologies such as seawater irrigation and seawater crop cultivation can also be employed in agricultural drought prevention and control [51,59], and can also be a topic worthy of in-depth research.

## 5. Conclusions

The present study determined variations and the trends of SPEI throughout the LMRB during 1980–2018 in the rainy season, dry season, and annual scale. The annual variations of SPEI were related to four atmospheric circulation factors. The major conclusions were:

(1) During the research period (1980–2018), the LMRB experienced frequent dry and wet changes on a monthly and seasonal scale. November to April of the following year is a period prone to meteorological drought. Especially after 1990, the distinction between dry and wet years became more pronounced. The risk of meteorological drought events increased annually, especially in 2015 with greater intensity than in previous years, posing a significant threat to the basin.

(2) The LMRB has well-defined dry and wet seasons, with the southwest monsoon bringing humidity and rainfall from May until the end of September. However,

meteorological drought events continue to persist, particularly in the western region of the Khorat Plateau, the Tonle Sap Lake Basin, where a drought trend has been observed. These areas correspond to the main agricultural areas in the basin. The frequency of severe and extreme meteorological droughts during the dry season has significantly risen, particularly in the upper reaches of the Lancang River and western Khorat Plateau. In comparison to other regions, this area experiences a higher frequency of severe meteorological droughts, accounting for approximately 10% of occurrences.

(3) There are differences in the response relationship between meteorological drought events in the LMRB and different atmospheric circulations (ENSO, AO, NAO, and PAO), with the strongest response being observed between the ENSO and meteorological drought events in the basin. Specifically, meteorological drought events occurring in the Northern Plateau, western parts of the Khorat Plateau, western parts of the Tonle Sap Lake Basin, and the Mekong Delta exhibit synchronicity with the ENSO. Meteorological drought events in the lower regions of the Lancang River basin show synchronicity with the AO, and there is a significant correlation between meteorological drought events in the Long Mountains and a lag of 2 months in the AO.

This study provides valuable insights for predicting meteorological drought in the LMRB by analyzing the teleconnections between atmospheric circulation and drought. Especially, the areas prone to meteorological drought and high-risk areas in the LMRB coincide with the main agricultural areas in the basin, increasing the risk of agricultural planting and production activities. Based on the above results, we can monitor the ENSO, AO, NAO, and PDO and their synchronicity or lag with meteorological drought, intervene and prevent meteorological drought events in the basin in advance, and provide beneficial scientific support for effective water resource management in the basin.

**Author Contributions:** Conceptualization, L.F. and Y.W.; methodology, L.F.; validation, L.F. and Y.W.; formal analysis, L.F.; investigation, L.F.; resources, L.F.; data curation, L.F.; writing—original draft preparation, L.F. and Y.W.; writing—review and editing, L.F., Y.W. and C.C.; project administration, Y.W. and W.C. All authors have read and agreed to the published version of the manuscript.

**Funding:** This research was funded by the Jiangsu Province Natural resources Science and Technology Project, grant number 2022016; Jiangsu Natural Resources Development Special Fund for Marine Science and Technology Innovation, grant number JSZRHYKJ202205; the National Natural Science Foundation of China, grant number 42177065; the Key R&D Project of Guangdong Province, grant number 2020B1111530001; and Guangdong Foundation for Program of Science and Technology Research, grant numbers 2019B121201004, 2019QN01L682, the National Key Research and Development Project, grant number 2023YFC3205701, and the GDAS Special Project of Science and Technology Development, grant number 2020GDASYL-20200102013.

**Institutional Review Board Statement:** Not applicable.

**Informed Consent Statement:** Not applicable.

**Data Availability Statement:** Publicly available datasets were analyzed in this study. This data can be found here: Climatic Research Unit TS v.4.03 (CRU) database (https://crudata.uea.ac.uk/cru/data/hrg/cru_ts_4.03/ accessed on 1 January 2021), the international disaster database (https://public.emdat.be/data/ accessed on 1 October 2023).

**Acknowledgments:** We are thankful to authors for their helpful discussions and feedback throughout the research process. We extend our appreciation to the anonymous reviewers for their constructive comments that greatly improved the quality of this manuscript.

**Conflicts of Interest:** The authors declare no conflict of interest.

## References

1. Ayugi, B.; Tan, G.; Niu, R.; Dong, Z.; Ojara, M.; Mumo, L.; Babaousmail, H.; Ongoma, V. Evaluation of Meteorological Drought and Flood Scenarios over Kenya, East Africa. *Atmosphere* **2020**, *11*, 307. [CrossRef]

2. Wang, F.; Lai, H.; Li, Y.; Feng, K.; Zhang, Z.; Tian, Q.; Zhu, X.; Yang, H. Dynamic variation of meteorological drought and its relationships with agricultural drought across China. *Agric. Water Manag.* **2022**, *261*, 107301. [CrossRef]
3. Tan, Y.X.; Ng, J.L.; Huang, Y.F. Quantitative analysis of input data uncertainty for SPI and SPEI in Peninsular Malaysia based on the bootstrap method. *Hydrol. Sci. J.* **2023**, *68*, 1724–1737. [CrossRef]
4. Li, Q.; Zeng, T.; Chen, Q.; Han, X.; Weng, X.; He, P.; Zhou, Z.; Du, Y. Spatio-temporal changes in daily extreme precipitation for the Lancang–Mekong River Basin. *Nat. Hazards* **2022**, *115*, 641–672. [CrossRef]
5. Lehner, B.; Döll, P.; Alcamo, J.; Henrichs, T.; Kaspar, F. Estimating the Impact of Global Change on Flood and Drought Risks in Europe: A Continental, Integrated Analysis. *Clim. Chang.* **2006**, *75*, 273–299. [CrossRef]
6. Stagge, J.H.; Kohn, I.; Tallaksen, L.M.; Stahl, K. Modeling drought impact occurrence based on meteorological drought indices in Europe. *J. Hydrol.* **2015**, *530*, 37–50. [CrossRef]
7. Luo, X.; Luo, X.; Ji, X.; Ming, W.; Wang, L.; Xiao, X.; Xu, J.; Liu, Y.; Li, Y. Meteorological and hydrological droughts in the Lancang-Mekong River Basin: Spatiotemporal patterns and propagation. *Atmos. Res.* **2023**, *293*, 106913. [CrossRef]
8. Hao, Z.; Singh, V.P. Drought characterization from a multivariate perspective: A review. *J. Hydrol.* **2015**, *527*, 668–678. [CrossRef]
9. Won, J.; Choi, J.; Lee, O.; Kim, S. Copula-based Joint Drought Index using SPI and EDDI and its application to climate change. *Sci. Total Environ.* **2020**, *744*, 140701. [CrossRef]
10. Zhao, H.; Gao, G.; An, W.; Zou, X.; Li, H.; Hou, M. Timescale differences between SC-PDSI and SPEI for drought monitoring in China. *Phys. Chem. Earth Parts A/B/C* **2017**, *102*, 48–58. [CrossRef]
11. Stagge, J.H.; Tallaksen, L.M.; Gudmundsson, L.; Van Loon, A.F.; Stahl, K. Candidate Distributions for Climatological Drought Indices (SPI and SPEI). *Int. J. Climatol.* **2015**, *35*, 4027–4040. [CrossRef]
12. Spinoni, J.; Barbosa, P.; De Jager, A.; McCormick, N.; Naumann, G.; Vogt, J.V.; Magni, D.; Masante, D.; Mazzeschi, M. A new global database of meteorological drought events from 1951 to 2016. *J. Hydrol. Reg. Stud.* **2019**, *22*, 100593. [CrossRef] [PubMed]
13. Ling, M.; Han, H.; Hu, X.; Xia, Q.; Guo, X. Drought characteristics and causes during summer maize growth period on Huang-Huai-Hai Plain based on daily scale SPEI. *Agric. Water Manag.* **2023**, *280*, 108198. [CrossRef]
14. Li, L.; She, D.; Zheng, H.; Lin, P.; Yang, Z.-L. Elucidating Diverse Drought Characteristics from Two Meteorological Drought Indices (SPI and SPEI) in China. *J. Hydrometeorol.* **2020**, *21*, 1513–1530. [CrossRef]
15. Mohammed, S.; Alsafadi, K.; Enaruvbe, G.O.; Bashir, B.; Elbeltagi, A.; Szeles, A.; Alsalman, A.; Harsanyi, E. Assessing the impacts of agricultural drought (SPI/SPEI) on maize and wheat yields across Hungary. *Sci. Rep.* **2022**, *12*, 8838. [CrossRef] [PubMed]
16. Manzano, A.; Clemente, M.A.; Morata, A.; Luna, M.Y.; Beguería, S.; Vicente-Serrano, S.M.; Martín, M.L. Analysis of the atmospheric circulation pattern effects over SPEI drought index in Spain. *Atmos. Res.* **2019**, *230*, 104630. [CrossRef]
17. Wang, W.; Zhu, Y.; Xu, R.; Liu, J. Drought severity change in China during 1961–2012 indicated by SPI and SPEI. *Nat. Hazards* **2014**, *75*, 2437–2451. [CrossRef]
18. Meque, A.; Abiodun, B.J. Simulating the link between ENSO and summer drought in Southern Africa using regional climate models. *Clim. Dyn.* **2014**, *44*, 1881–1900. [CrossRef]
19. Zhou, L.; Wang, S.; Du, M.; Chen, Q.; He, C.; Zhang, J.; Zhu, Y.; Gong, Y. The Influence of ENSO and MJO on Drought in Different Ecological Geographic Regions in China. *Remote Sens.* **2021**, *13*, 875. [CrossRef]
20. Chandrasekara, S.S.K.; Kwon, H.-H.; Vithanage, M.; Obeysekera, J.; Kim, T.-W. Drought in South Asia: A Review of Drought Assessment and Prediction in South Asian Countries. *Atmosphere* **2021**, *12*, 369. [CrossRef]
21. Xing, Z.; Yu, Z.; Wei, J.; Zhang, X.; Ma, M.; Yi, P.; Ju, Q.; Wang, J.; Laux, P.; Kunstmann, H. Lagged influence of ENSO regimes on droughts over the Poyang Lake basin, China. *Atmos. Res.* **2022**, *275*, 106218. [CrossRef]
22. Rezaei, A. Ocean-atmosphere circulation controls on integrated meteorological and agricultural drought over Iran. *J. Hydrol.* **2021**, *603*, 126928. [CrossRef]
23. Li, X.; Sha, J.; Wang, Z.-L. Influence of the Three Gorges Reservoir on climate drought in the Yangtze River Basin. *Environ. Sci. Pollut. Res.* **2021**, *28*, 29755–29772. [CrossRef] [PubMed]
24. Lv, A.; Fan, L.; Zhang, W. Impact of ENSO Events on Droughts in China. *Atmosphere* **2022**, *13*, 1764. [CrossRef]
25. Zhang, B.; Li, Y.; Zhang, C.; Hu, C.; Fu, G.; Cai, X. Dual water-electricity cooperation improves economic benefits and water equality in the Lancang-Mekong River Basin. *Nat. Commun.* **2023**, *14*, 6228. [CrossRef] [PubMed]
26. Vicente-Serrano, S.M.; Beguería, S.; López-Moreno, J.I. A Multiscalar Drought Index Sensitive to Global Warming: The Standardized Precipitation Evapotranspiration Index. *J. Clim.* **2010**, *23*, 1696–1718. [CrossRef]
27. Sun, C.; Xiao, Z.-N.; Nguyen, M. Projection on precipitation frequency of different intensities and precipitation amount in the Lancang-Mekong River basin in the 21st century. *Adv. Clim. Chang. Res.* **2021**, *12*, 162–171. [CrossRef]
28. Do, P.; Tian, F.; Zhu, T.; Zohidov, B.; Ni, G.; Lu, H.; Liu, H. Exploring synergies in the water-food-energy nexus by using an integrated hydro-economic optimization model for the Lancang-Mekong River basin. *Sci. Total Environ.* **2020**, *728*, 137996. [CrossRef]
29. Luo, X.; Wang, Y.; Li, Y. Responses of ecosystem water use efficiency to drought in the Lancang–Mekong River Basin. *Front. Ecol. Evol.* **2023**, *11*, 1203725. [CrossRef]
30. Ty, T.V.; Lavane, K.; Nguyen, P.C.; Downes, N.K.; Nam, N.D.G.; Minh, H.V.T.; Kumar, P. Assessment of Relationship between Climate Change, Drought, and Land Use and Land Cover Changes in a Semi-Mountainous Area of the Vietnamese Mekong Delta. *Land* **2022**, *11*, 2175. [CrossRef]

31. Kang, H.; Sridhar, V.; Ali, S.A. Climate change impacts on conventional and flash droughts in the Mekong River Basin. *Sci. Total Environ.* **2022**, *838*, 155845. [CrossRef]
32. Ming, W.; Luo, X.; Luo, X.; Long, Y.; Xiao, X.; Ji, X.; Li, Y. Quantitative Assessment of Cropland Exposure to Agricultural Drought in the Greater Mekong Subregion. *Remote Sens.* **2023**, *15*, 2737. [CrossRef]
33. Han, X.; Li, Q.; Yang, X.; Xu, S.; Zou, Z.; Deng, M.; Wang, W. The influence of anthropogenic climate change on meteorological drought in the Lancang-Mekong River basin. *J. Hydrol.* **2023**, *626*, 130334. [CrossRef]
34. Wang, S.; Zhang, L.; She, D.; Wang, G.; Zhang, Q. Future projections of flooding characteristics in the Lancang-Mekong River Basin under climate change. *J. Hydrol.* **2021**, *602*, 126778. [CrossRef]
35. Ullah, S.; You, Q.; Sachindra, D.A.; Nowosad, M.; Ullah, W.; Bhatti, A.S.; Jin, Z.; Ali, A. Spatiotemporal changes in global aridity in terms of multiple aridity indices: An assessment based on the CRU data. *Atmos. Res.* **2022**, *268*, 105998. [CrossRef]
36. Beguería, S.; Vicente-Serrano, S.M.; Reig, F.; Latorre, B. Standardized precipitation evapotranspiration index (SPEI) revisited: Parameter fitting, evapotranspiration models, tools, datasets and drought monitoring. *Int. J. Climatol.* **2014**, *34*, 3001–3023. [CrossRef]
37. Vicente-Serrano, S.M.; López-Moreno, J.I.; Gimeno, L.; Nieto, R.; Morán-Tejeda, E.; Lorenzo-Lacruz, J.; Beguería, S.; Azorin-Molina, C. A multiscalar global evaluation of the impact of ENSO on droughts. *J. Geophys. Res.* **2011**, *23*, 1696. [CrossRef]
38. Luo, N.; Mao, D.; Wen, B.; Liu, X. Climate Change Affected Vegetation Dynamics in the Northern Xinjiang of China: Evaluation by SPEI and NDVI. *Land* **2020**, *9*, 90. [CrossRef]
39. Wu, R.; Zhang, J.; Bao, Y.; Guo, E. Run Theory and Copula-Based Drought Risk Analysis for Songnen Grassland in Northeastern China. *Sustainability* **2019**, *11*, 6032. [CrossRef]
40. Pei, Z.; Fang, S.; Wang, L.; Yang, W. Comparative Analysis of Drought Indicated by the SPI and SPEI at Various Timescales in Inner Mongolia, China. *Water* **2020**, *12*, 1925. [CrossRef]
41. Moradi, H.R.; Rajabi, M.; Faragzadeh, M. Investigation of meteorological drought characteristics in Fars province, Iran. *Catena* **2011**, *84*, 35–46. [CrossRef]
42. Senatilleke, U.; Sirisena, J.; Gunathilake, M.B.; Muttil, N.; Rathnayake, U. Monitoring the Meteorological and Hydrological Droughts in the Largest River Basin (Mahaweli River) in Sri Lanka. *Climate* **2023**, *11*, 57. [CrossRef]
43. Cao, S.; Zhang, L.; He, Y.; Zhang, Y.; Chen, Y.; Yao, S.; Yang, W.; Sun, Q. Effects and contributions of meteorological drought on agricultural drought under different climatic zones and vegetation types in Northwest China. *Sci. Total Environ.* **2022**, *821*, 153270. [CrossRef] [PubMed]
44. Sharma, A.; Goyal, M.K. Assessment of drought trend and variability in India using wavelet transform. *Hydrol. Sci. J.* **2020**, *65*, 1539–1554. [CrossRef]
45. Kingston, D.G.; Stagge, J.H.; Tallaksen, L.M.; Hannah, D.M. European-Scale Drought: Understanding Connections between Atmospheric Circulation and Meteorological Drought Indices. *J. Clim.* **2015**, *28*, 505–516. [CrossRef]
46. Rahmani, F.; Fattahi, M.H. A multifractal cross-correlation investigation into sensitivity and dependence of meteorological and hydrological droughts on precipitation and temperature. *Nat. Hazards* **2021**, *109*, 2197–2219. [CrossRef]
47. Zhang, R.; Chen, Z.Y.; Xu, L.J.; Ou, C.Q. Meteorological drought forecasting based on a statistical model with machine learning techniques in Shaanxi province, China. *Sci. Total Environ.* **2019**, *665*, 338–346. [CrossRef]
48. Dong, Z.; Liu, H.; Baiyinbaoligao; Hu, H.; Khan, M.Y.A.; Wen, J.; Chen, L.; Tian, F. Future projection of seasonal drought characteristics using CMIP6 in the Lancang-Mekong River Basin. *J. Hydrol.* **2022**, *610*, 127815. [CrossRef]
49. Yun, X.; Tang, Q.; Li, J.; Lu, H.; Zhang, L.; Chen, D. Can reservoir regulation mitigate future climate change induced hydrological extremes in the Lancang-Mekong River Basin? *Sci. Total Environ.* **2021**, *785*, 147322. [CrossRef]
50. Li, D.; Zhao, J.; Govindaraju, R.S. Water benefits sharing under transboundary cooperation in the Lancang-Mekong River Basin. *J. Hydrol.* **2019**, *577*, 123989. [CrossRef]
51. Minh, H.V.T.; Lavane, K.; Ty, T.V.; Downes, N.K.; Hong, T.T.K.; Kumar, P. Evaluation of the Impact of Drought and Saline Water Intrusion on Rice Yields in the Mekong Delta, Vietnam. *Water* **2022**, *14*, 3499. [CrossRef]
52. Nguyen, T.T.; Li, M.H.; Vu, T.M.; Chen, P.Y. Multiple drought indices and their teleconnections with ENSO in various spatiotemporal scales over the Mekong River Basin. *Sci. Total Environ.* **2023**, *854*, 158589. [CrossRef]
53. Zhang, Y.; Hao, Z.; Feng, S.; Zhang, X.; Xu, Y.; Hao, F. Agricultural drought prediction in China based on drought propagation and large-scale drivers. *Agric. Water Manag.* **2021**, *255*, 107028. [CrossRef]
54. Feng, P.; Wang, B.; Luo, J.J.; Liu, L.; Waters, C.; Ji, F.; Ruan, H.; Xiao, D.; Shi, L.; Yu, Q. Using large-scale climate drivers to forecast meteorological drought condition in growing season across the Australian wheatbelt. *Sci. Total Environ.* **2020**, *724*, 138162. [CrossRef] [PubMed]
55. Venkatappa, M.; Sasaki, N.; Han, P.; Abe, I. Impacts of droughts and floods on croplands and crop production in Southeast Asia—An application of Google Earth Engine. *Sci. Total Environ.* **2021**, *795*, 148829. [CrossRef] [PubMed]
56. Minh, H.V.T.; Van Ty, T.; Avtar, R.; Kumar, P.; Le, K.N.; Ngan, N.V.C.; Khanh, L.H.; Nguyen, N.C.; Downes, N.K. Implications of climate change and drought on water requirements in a semi-mountainous region of the Vietnamese Mekong Delta. *Environ. Monit. Assess.* **2022**, *194*, 766. [CrossRef] [PubMed]
57. Liu, J.; Chen, D.; Mao, G.; Irannezhad, M.; Pokhrel, Y. Past and Future Changes in Climate and Water Resources in the Lancang–Mekong River Basin: Current Understanding and Future Research Directions. *Engineering* **2022**, *13*, 144–152. [CrossRef]

58. Zhang, L.; Song, W.; Song, W. Assessment of Agricultural Drought Risk in the Lancang-Mekong Region, South East Asia. *Int. J. Environ. Res. Public Health* **2020**, *17*, 6153. [CrossRef] [PubMed]
59. Lavane, K.; Kumar, P.; Meraj, G.; Han, T.G.; Ngan, L.H.B.; Lien, B.T.B.; Van Ty, T.; Thanh, N.T.; Downes, N.K.; Nam, N.D.G.; et al. Assessing the Effects of Drought on Rice Yields in the Mekong Delta. *Climate* **2023**, *11*, 13. [CrossRef]

*atmosphere*

*Article*

# Application of an Ensemble Stationary-Based Category-Based Scoring Support Vector Regression to Improve Drought Prediction in the Upper Colorado River Basin

**Mohammad Hadi Bazrkar [1,*], Heechan Han [2], Tadesse Abitew [3], Seonggyu Park [1], Negin Zamani [4] and Jaehak Jeong [1]**

[1] Blackland Research and Extension Center, Texas A&M AgriLife Research, Temple, TX 76502, USA; seonggyu.park@brc.tamus.edu (S.P.); jaehak.jeong@brc.tamus.edu (J.J.)
[2] Department of Civil Engineering, Chosun University, Gwangju 61452, Republic of Korea; heechan@chosun.ac.kr
[3] Texas Water Development Board, 1700 N. Congress Ave, 5th FL, P.O. Box 13231, Austin, TX 78711, USA; tadesse.abitew@twdb.texas.gov
[4] Department of Civil and Architectural Engineering, Texas A&M University Kingsville, Kingsville, TX 78363, USA; negin.zamani@students.tamuk.edu
* Correspondence: hadi.bazrkar@ag.tamu.edu

**Abstract:** Recent above-normal temperatures, which exacerbated the impacts of precipitation deficits, are recognized as the primary driver of droughts in the Upper Colorado River Basin (UCRB), USA. This research aims to enhance drought prediction models by addressing structural changes in non-stationary temperature time series and minimizing drought misclassification through the ES-CBS-SVR model, which integrates ESSVR and CBS-SVR. The research investigates whether this coupling improves prediction accuracy. Furthermore, the model's performance will be tested in a region distinct from those originally used to evaluate its generalizability and effectiveness in forecasting drought conditions. We used a change point detection technique to divide the non-stationary time series into stationary subsets. To minimize the chances of drought mis-categorization, category-based scoring was used in ES-CBS-SVR. In this study, we tested and compared the ES-CBS-SVR and SVR models in the Upper Colorado River Basin (UCRB) using data from the Global Land Data Assimilation System (GLDAS), where the periods 1950–2004 and 2005–2014 were used for training and testing, respectively. The results indicated that ES-CBS-SVR outperformed SVR consistently across of the drought indices used in this study in a higher portion of the UCRB. This is mainly attributed to variable hyperparameters (regularization constant and tube size) used in ES-CBS-SVR to deal with structural changes in the data. Overall, our analysis demonstrated that the ES-CBS-SVR can predict drought more accurately than traditional SVR in a warming climate.

**Keywords:** Upper Colorado River Basin; drought; identification; categorization; prediction; non-stationarity

## 1. Introduction

The Colorado River Basin, the most overallocated basin in the world [1], provides water to about 40 million people in the southwest United States and Mexico [2] and water for irrigation of 2.2 million hectares of land [3]. Despite the critical importance of the basin, the scarce water resources have faced big challenges due to climate change [4] and anthropogenic factors [2,5], Many climate change scenarios involve decreases in the late-summer precipitation and mean annual streamflow [6] and thus more severe droughts [7] and more wildfires in the Upper Colorado River Basin (UCRB) [2]. The most extreme low-frequency drought occurred in the mid-1100s, characterized by a decrease in mean annual flow and the absence of high annual flows over six decades [8,9]. The drought from 1923 to 2004 is the worst in the recorded period, ranking 6th to 14th in magnitude and 1st to 12th in severity in the past 500 years [10]. The current millennium drought (2000–2022) has an average flow far lower than the historical record [11]. Thus, it is critical to improve identification, categorization, and prediction of drought in the UCRB.

Snow, snowpack, and snowmelt play essential roles in the UCRB's hydrologic system. McCabe et al. (2020) [12] demonstrated that winter precipitation deficits are the primary drivers of droughts in the UCRB rather than warm temperature anomalies. Eight drought periods were identified during the 1901–2014 period. The driest drought period spanned 1901–1904, whereas the longest drought period occurred between 1943 and 1956 [12]. Miller et al. (2011) [13] indicated that the timing of the last day of the snow season corresponds well to the volume of runoff observed over the traditional peak flow season (April through July); conversely, the timing of the first day of the snow season does not correspond well to the volume of runoff observed over the peak flow season [13]. Standardized precipitation and evapotranspiration index (SPEI; [14]) and standardized precipitation index (SPI; [15]) in the UCRB show similar temporal and spatial patterns, but the inclusion of temperatures in SPEI leads to more extreme magnitudes in SPEI than in SPI [16]. Thus, it is essential to consider temperature and snow-based drought indices in drought analysis in the UCRB.

Increasing temperatures, combined with moderate precipitation deficits and soil moisture deficits, are causing amplifications in recent droughts. This has resulted in marked increases in warm years with lower flows than expected, given the precipitation received. A simple statistical model of water year streamflow with temperatures increased by 1–4 °C in the UCRB showed reductions in flow and runoff efficiency with each degree of warming [4,17]. Estimates of Colorado River flow sensitivity to temperature, along with climate model-based temperature projections, indicate that continued warming will lead to temperature-induced declines in river flow, ranging from −20% by mid-century to −55% by the end of the century [18]. The more significant effect of warm-season temperature (April to September) on the variability of UCRB flow suggests that evaporation or snowmelt has driven recent reductions in UCRB flow [19]. The model simulations indicate that soil moisture deficits could persist for 12 years or more in the future, and sustained water supplies in parts of the UCRB will be a challenge as the climate continues to warm [20]. Soil moisture anomalies were lower than −10 mm in 15 out of 19 drought years [21].

Drought prediction in the UCRB is of interest to the researchers. Mohsin and Pilz (2021) [22] reported a reliable forecast of the interarrival time of drought by using explicit distribution based on the convolution of stochastic variables derived from the Bivariate Affine-Linear Exponential (BALE) distribution. Madadgar and Moradkhani (2013) [23] used the standardized streamflow index [24] in a multivariate probabilistic framework to forecast spring flow (April–June) as a representative of hydrologic droughts with different severities in the UCRB. The spring flow showed higher correlations with the previous winter (January–March) than with fall (October–December). Two major predictors, teleconnection indices and landscape variables, have been used for drought prediction in the UCRB. Teleconnection–drought relationships are not highly correlated for the UCRB compared with the southern portion of the Colorado River Basin [25]. Conversely, applying landscape variables (e.g., precipitation, percent snow, potential evapotranspiration, soils, and drainage area) for predicting low flow metrics has been more successful [6].

With the hydroclimatic changes due to climate change and/or variability, the validity of historical stationarity assumptions in the hydroclimatic variables has been increasingly questioned [26,27]. The drought prediction models are based on the stationarity assumption that the generating process is in equilibrium around an underlying mean and that variance remains constant over time [27]. Thus, the performances of drought prediction models may not be satisfactory if this stationary assumption is not satisfied [28,29]. It is critical to consider non-stationarity in drought prediction models [30,31]. Non-stationary and statistically significant increases in temperature time series have been reported in the UCRB. Although Murphy and Ellis (2014) [27] indicated no persistent reduction and non-stationary in precipitation and runoff time series in the UCRB, non-stationary in streamflow over decadal and longer timescales is frequently reported in other studies. Talsma et al. (2022) [32] applied an unsupervised machine learning based on Non-Negative Matrix Factorization using K-means clustering (NMFk) to efficiently identify behavioral changes in drought indicators such as snowpack, snowmelt timing, precipitation, and evapotranspiration across space and time and to quickly analyze and interpret hydro climate model results [32]. Significant shifts in peak runoff have been reported in snowmelt-dominant sub-watersheds, with the complete disappearance of the snowmelt signal for some sub-watersheds. This significant non-stationarity makes short-term records inappropriate for most planning and forecast applications with critical implications for the UCRB water resource management [5,33].

Change point detection methods have been increasingly used in drought studies to overcome the above-mentioned non-stationarity problem in different models [30,34,35]. Bazrkar and Chu (2021) [26] developed a new ensemble stationary-based support vector regression (ESSVR) method to improve the prediction of droughts in a changing climate. A change point detection technique has been specifically coupled with the widely used SVR to enhance its performance for drought prediction.

Accurate predictions of drought categories are crucial for stakeholders to decide how to prepare for potential droughts. However, using the existing standard performance metrics such as RMSE for training SVR-based drought prediction models potentially leads to the mis-categorization of droughts since such metrics minimize the difference between the actual and predicted values by only accounting for this numerical difference and not categorical discrepancies. Bazrkar and Chu (2022) [36] developed a category-based scoring SVR (CBS-SVR) method compared with the traditional SVR and support vector classification (SVC) since these models share some similarities in their concepts. SVR is a numerical model with the potential for mis-categorization.

The objective of this research is to improve drought prediction in the UCRB by developing an ensemble stationery-based and category-based scoring support vector regression (ES-CBS-SVR), which is a coupled application of ESSVR [26] and CBS-SVR [36]. The research explores whether coupling these models enhances drought prediction accuracy. Additionally, the model's effectiveness can be further evaluated by applying it to other regions to assess its capability in forecasting drought under varying climatic and hydrologic conditions. This paper is organized as follows: Section 2 describes the methodology, algorithm and framework of the ES-CBS-SVR model and evaluation of ES-CBS-SVR. Section 3 provides the results of ES-CBS-SVR and traditional SVR and their comparisons. Section 4 discusses the results. Finally, Section 5 presents the conclusions.

## 2. Methodology

The proposed ES-CBS-SVR method combines ESSVR [26] and CBS-SVR [36] to enhance drought prediction. ESSVR addresses non-stationarity in temperature time series by applying change point detection to divide the data into stationary subsets. SVR models are trained on each subset, and their predictions are weighted to improve accuracy. CBS-SVR introduces a category-based scoring approach to optimize hyperparameters, reducing misclassification. By integrating these models, the study aims to minimize the risk of failure in drought prediction. Figure 1 illustrates the flowchart detailing the approach for developing ES-CBS-SVR.

### 2.1. Study Area

The UCRB drains an area of more than 279,720 km$^2$ in seven states (Figure 2) and generates water for 26 million people within the basin states and the adjoining regions. Approximately 64% of the basin is considered arid or semi-arid, with annual precipitation of 370 mm and an average yearly temperature of 6 °C [13]. Elevation ranges from 900 m in the southwest to 4300 m in the northeast. The UCRB land cover is dominated by rangeland (65%) and evergreen forest (25%). One of the prominent droughts in the UCRB occurred in 2000 and resulted in an accumulative streamflow deficit of 11 km$^3$ or approximately two years of average streamflow [21]. Droughts combined with an increased water supply demand have severely affected the storage of the major reservoirs at the Colorado River

Basin [21]. To lessen the effects of drought and increase water storage for use during dry periods, several dams were constructed on the mainstem and tributaries of the Colorado River upstream of Lake Buchanan since the late 1940s. Analysis of flow at the gauge above Lake Buchanan indicated streamflow was significantly reduced during the recent drought (2009–2014), compared with streamflows during the 'drought of record' (1950–1957) [37]. Temperature also plays a critical role in droughts and wet conditions on Colorado River flows [38].

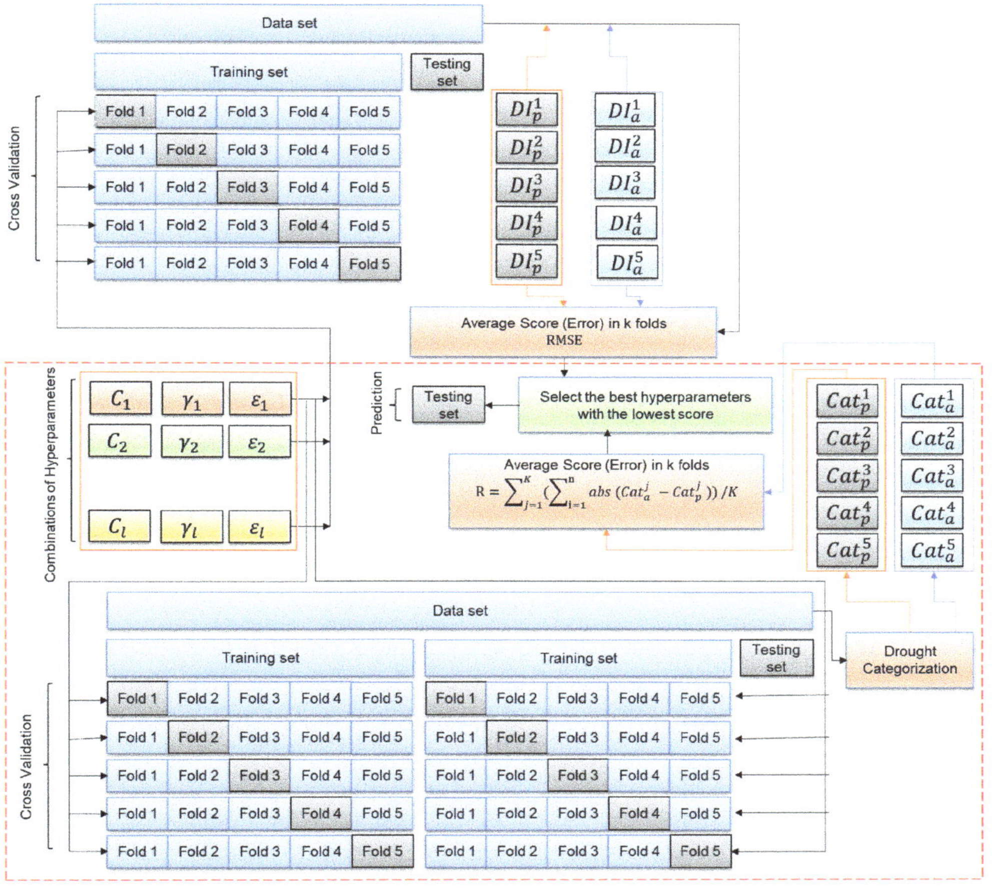


**Figure 1.** Cross-validation process in the ES-CBS-SVR model (adapted from [26,36]).

On average, 56% of the streamflow in the UCRB originated as base flow, and precipitation was identified as the dominant driver of spatial variability in base flow at the scale of the UCRB, with the majority of base flow discharge to streams occurring in upper elevation watersheds. The model estimates an average of $1.8 \times 10^{10}$ $m^3/y$ of base flow in the UCRB, greater than 80% of which is lost during in-stream transport to the Lower Colorado River Basin via processes including evapotranspiration and water diversion for irrigation. Miller et al. (2016) [13] indicated that surface waters in the Colorado River Basin are dependent on base flow and that management approaches that consider groundwater and surface water as joint resources will be needed to effectively manage current and future water resources in the Basin.

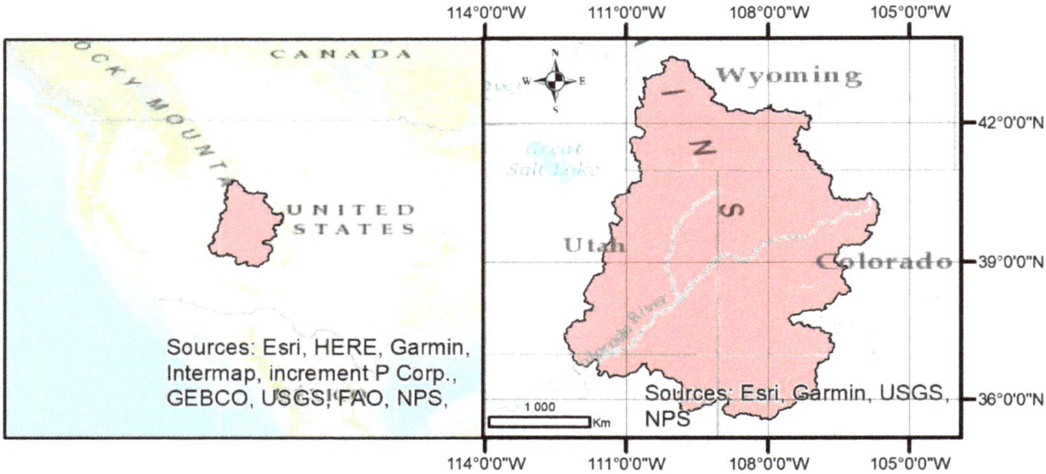


**Figure 2.** Location of the Upper Colorado River Basin (UCRB).

### 2.2. *ES-CBS-SVR Model*

A fast-approximate window-based change point detection method, window sliding [39], is used to split non-stationary time series into multiple stationary time series. SVR is further conducted on each stationary subset of the training data. An initial weight is assigned to the predicted values by SVR on each subset of data, and it is updated by comparison with the actual values and higher weights assigned to those with higher accuracy. Finally, the prediction with the higher weight is selected as the final prediction using category-based scoring [26].

A fast, approximate window-based change point detection method, window sliding [39], is employed to partition non-stationary time series into stationary segments. SVR is applied to each stationary subset, with initial weights assigned to predicted values and updated based on accuracy compared to actual data. Higher weights are given to more accurate predictions, and the final prediction is selected using category-based scoring.

SVR, originally proposed by [40], models relationships between predictors and predictands. Bazrkar and Chu (2021) [26] enhanced it to address non-stationarity in temperature datasets for drought prediction in a warming climate. Hyperparameter tuning, critical to model performance, involves optimizing Cost (C), epsilon ($\varepsilon$), and gamma (G) using grid search cross-validation within each stationary subset. Predictions closest to actual values (based on RMSE) are selected, with an ensemble technique applied to combine outputs.

Traditional SVR uses numerical scoring (e.g., RMSE), but ES-CBS-SVR adopts category-based scoring, reducing misclassification risk. The risk scoring approach [36] compares actual and predicted categories, selecting the lowest-risk prediction for optimal model performance.

### 2.3. *Drought Identification*

The target variables include multivariate, bivariate, and univariate standardized drought indices. The multivariate Hydroclimatic Aggregate Drought Index (HADI) [41] is calculated using principal component analysis (PCA) on rainfall, snowmelt, surface runoff, and soil moisture. Bivariate indices, such as SPEI and SMRI, are derived from differences in precipitation-evapotranspiration and the sum of snowmelt and rainfall, respectively. Univariate indices, including SPI, SRI, and SSI, standardize cumulative distribution functions (CDFs) of precipitation, runoff, and soil moisture, using Gringorten plotting positions for consistency with methods by [42]).

### 2.4. Drought Categorization

Drought categorization is crucial in this study as it influences SVR hyperparameters and affects drought predictions. Traditional fixed thresholds often fail to capture spatial and temporal variations, prompting the adoption of a Customized Drought Categorization (CDC) proposed by Bazrkar et al. (2020) [41]. The CDC employs cell-by-cell analysis and uses joint probability distribution and conditional expectation to determine average drought category probabilities. Variable thresholds are derived through K-means clustering, with categories aligned to [43], ranging from exceptional drought (D4) to exceptional wet conditions (W4). Due to the limited study period, extreme thresholds were predefined at −2 and 2.

### 2.5. Drought Prediction

To set up the ES-CBS-SVR model for drought prediction, three types of drought indices in monthly time scales are used. The target variables are six drought indices. Depending on the specific drought index that is used as the target variable in the prediction model, different types of predictor(s) are used. Precipitation (P) is a common predictor in all drought indices. Temperature (T) is also used for the prediction of SPEI and SMRI. In addition, actual evapotranspiration (ET) and DFCT are added to the prediction of HADI, SSI, and SRI.

The training and testing periods range from 1950 to 2004 and 2005 to 2014, respectively. The input data for calculating the drought indices and prediction model are obtained from the Global Land Data Assimilation System (GLDAS) [44]. The goal of GLDAS is to ingest satellite- and ground-based observational data products, using advanced land surface modeling and data assimilation techniques, to generate optimal fields of land surface states and fluxes [44]. The data are globally available at resolutions of 2.5 degrees to 1 km in near-real time from1948 to present simulations. The required data for this study were limited to 2014.

To evaluate the overall (areal average) performances of ES-CBS-SVR and the traditional SVR, we used the Mean absolute error (MAE), the mean square error (MSE), the Root mean square error (RMSE), and the coefficient of determination ($R^2$). These performance metrics were calculated for each model by comparing the predicted and actual (monitored) values. A comparison between two models was further conducted based on these metrics.

## 3. Results

### 3.1. Drought Identification and Categorization in the UCRB

SPI, SPEI, SMRI, SRI, SSI, and HADI are based on different hydroclimatic variables. Thus, each index provides information about different types of droughts and their severity (Table 1). The results for SPI in the UCRB show that exceptional droughts occurred in October 1952, September 1953, November 1964, November 1976, June 1978, May 1979, June 1980, and June 2013. The SPEI results indicated that exceptional drought occurred in June 1978, May 1979, June 1980, June 1985, June 1987, July 1993, July 1995, and July 2005. The SMRI and SRI identified similar exceptional drought episodes and covered a significant portion of the UCRB in October 1952, October 1964, and June 2013. The SSI results showed that the exceptional drought occurred in September 1956, August 1960, June 2002, July 2002, August 2002, August 2012, and September 2012. HADI identified October 1972, February 1979, September 1982, April 1993, and May 1995 as exceptional droughts.

Considering the abnormal dry, moderate, severe, extreme, and exceptional drought categories, the droughts were identified in different durations based on each drought index (Figures 3 and 4). According to SPI, the most prolonged dry period was six consecutive months, from January to February 2002 and May to June 2002. There were two two-month dry periods with the one-month normal condition between May to June 1974 and August to September 1974. The other long dry periods were identified from October to December 1976 and November 1980 to February 1981. The last relatively long dry period occurred in March 2012 and May to June 2012, with a one-month abnormal dry condition in between.

For the SPEI, the most severe and long dry period occurred from April to September 1979. Generally, based on SPEI, long-term droughts were identified in 1952, 1958, 1960, 1962, 1973, 1979, 1988, 1994, 1995, 1997, 1998, 2000, and 2005. According to the results for SMRI, there are relatively long dry periods for SRI compared with the other drought indices. The 1963–1964, 1973, 1976, 1999, 2002, and 2012 were dry years based on the SMRI. The SRI results indicated that 1956, 1974, 2002, and 2012 were dry years with the deficit in runoff and streamflow in the UCRB. Lack of soil moisture, which is reflected in the SSI, was observed in almost a significant portion of the 50s decade, including 1950, 1954 to 1956, 1958, 1964, 1974, 1976, 1977, 1989, 1994, 2002, 2003, 2012. The longest and the most severe deficit in soil moisture was associated with 1956 from June to December. The most extended dry period based on the HADI lasted three consequence months from April to June 2002. In this period, all the drought indices except SPEI showed dry conditions (Figure 4).

**Table 1.** Exceptional droughts based on SPI, SPEI, SMRI, SRI, SSI, and HADI.

| | SPI | SPEI | SMRI | SRI | SSI | HADI |
|---|---|---|---|---|---|---|
| Oct-52 | + | − | + | + | − | − |
| Sep-53 | + | − | − | − | − | − |
| Sep-56 | − | − | − | − | + | − |
| Aug-60 | − | − | − | − | + | − |
| Oct-64 | − | − | + | + | − | − |
| Nov-64 | + | − | − | − | − | − |
| Oct-72 | − | − | − | − | − | + |
| Nov-76 | + | − | − | − | − | − |
| Jun-78 | + | + | − | − | − | − |
| Feb-79 | − | − | − | − | − | + |
| May-79 | + | + | − | − | − | − |
| Jun-80 | + | + | − | − | − | − |
| Sep-82 | − | − | − | − | − | + |
| Jun-85 | − | + | − | − | − | − |
| Jun-87 | − | + | − | − | − | − |
| Apr-93 | − | − | − | − | − | + |
| Jul-93 | − | + | − | − | − | − |
| May-95 | − | − | − | − | − | + |
| Jul-95 | − | + | − | − | − | − |
| Jun-02 | − | − | − | − | + | − |
| Jul-02 | − | − | − | − | + | − |
| Aug-02 | − | − | − | − | + | − |
| Jul-05 | − | + | − | − | − | − |
| Aug-12 | − | − | − | − | + | − |
| Sep-12 | − | − | − | − | + | − |
| Jun-13 | + | − | + | + | − | − |

Note: "+" and "−" refer to occurrence and non-occurrence of drought by the specified drought index, respectively.

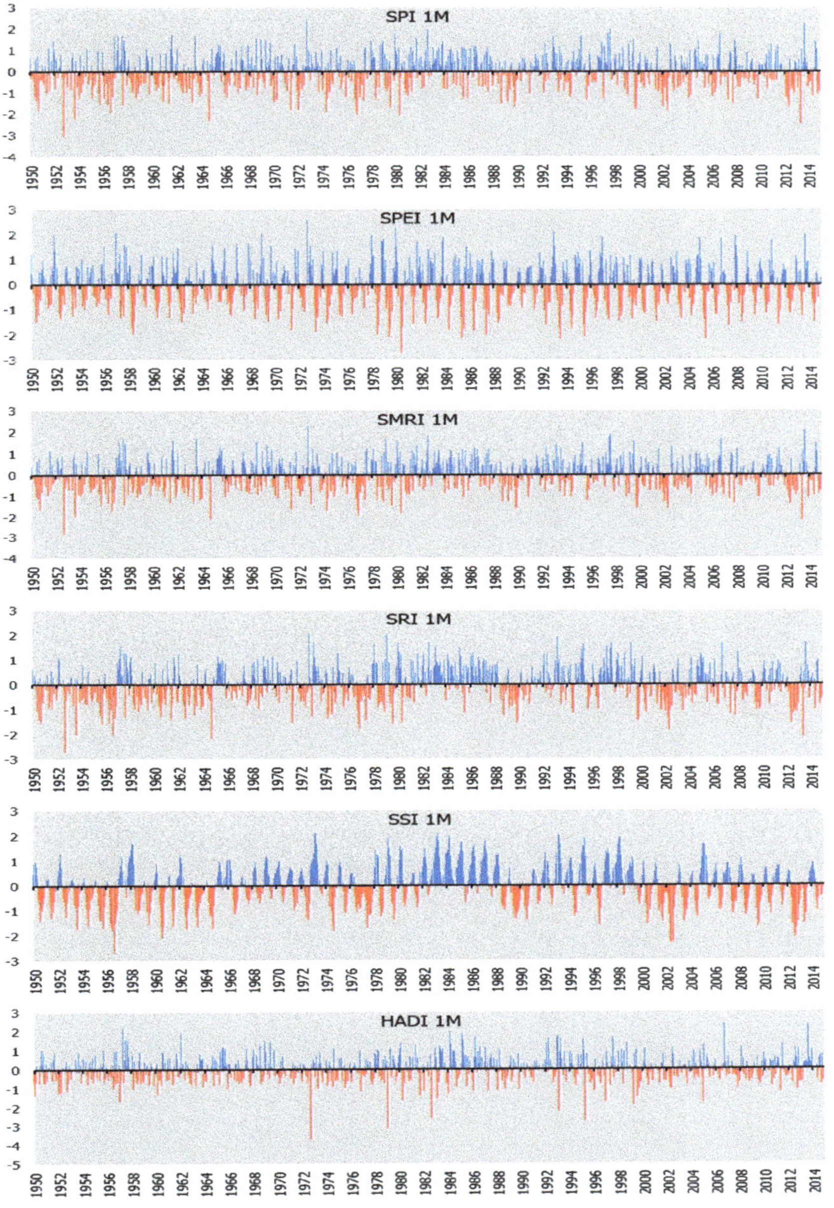


**Figure 3.** Drought identification in the UCRB (red and blue color show dry and wet conditions, respectively).

| M-Y | SPI | SPEI | SMRI | SRI | SSI | HADI |
|---|---|---|---|---|---|---|
| Aug-50 | | | | | -1.41 | |
| Sep-50 | | | | | -0.90 | |
| Oct-50 | | | | | -1.01 | |
| May-52 | | -1.12 | | | | |
| Jun-52 | | -1.33 | | | | |
| Jul-52 | | -1.02 | | | | |
| Aug-52 | | -0.40 | | | | |
| Sep-52 | | -0.42 | | | | |
| Oct-52 | | -0.79 | | | | |
| Jun-54 | | | | | -0.89 | |
| Jul-54 | | | | | -1.27 | |
| Aug-54 | | | | | -1.59 | |
| Sep-54 | | | | | -0.91 | |
| Jul-55 | | | | | -1.39 | |
| Aug-55 | | | | | -1.06 | |
| Sep-55 | | | | | -1.37 | |
| Oct-55 | | | | | -1.71 | |
| Jun-56 | | | | -1.50 | -1.46 | |
| Jul-56 | | | | -0.55 | -1.64 | |
| Aug-56 | | | | -0.97 | -1.76 | |
| Sep-56 | | | | -2.00 | -2.71 | |
| Oct-56 | | | | -0.44 | -1.51 | |
| Nov-56 | | | | -1.34 | -1.23 | |
| Dec-56 | | | | | -0.96 | |
| May-58 | | -1.71 | | | | |
| Jun-58 | | -1.97 | | | | |
| Jul-58 | | -1.46 | | | -0.97 | |
| Aug-58 | | | | | -1.13 | |
| Sep-58 | | | | | -1.02 | |
| Oct-58 | | | | | -0.98 | |
| May-60 | | -1.03 | | | | |
| Jun-60 | | -1.32 | | | | |
| Jul-60 | | -0.91 | | | | |
| May-62 | | -0.96 | | | | |
| Jun-62 | | -1.47 | | | | |
| Jul-62 | | -1.15 | | | | |
| Aug-62 | | -0.93 | | | | |
| Dec-63 | | | -1.16 | | | |
| Jan-64 | | | -1.06 | | | |
| Feb-64 | | | -1.10 | | | |
| Jul-64 | | | | | -1.16 | |
| Aug-64 | | | | | -1.17 | |
| Sep-64 | | | | | -1.22 | |
| Oct-64 | | | | | -1.73 | |

| M-Y | SPI | SPEI | SMRI | SRI | SSI | HADI |
|---|---|---|---|---|---|---|
| May-73 | | -0.94 | | | | |
| Jun-73 | | -1.89 | | | | |
| Jul-73 | | -1.24 | | | | |
| Aug-73 | | -1.33 | | | | |
| May-74 | -1.90 | | | -1.37 | | |
| Jun-74 | -1.26 | | | -1.13 | | |
| Jul-74 | 0.44 | | | -0.16 | -1.02 | |
| Aug-74 | -0.96 | | | -1.19 | -1.60 | |
| Sep-74 | -0.99 | | | -1.17 | -1.82 | |
| Oct-76 | -1.28 | | -1.12 | | -1.03 | |
| Nov-76 | -1.86 | | -1.69 | | -1.61 | |
| Dec-76 | -2.02 | | -1.94 | | -1.77 | |
| Jun-77 | | | | | -1.71 | |
| Jul-77 | | | | | -1.18 | |
| Aug-77 | | | | | -0.82 | |
| Sep-77 | | | | | -1.20 | |
| Oct-77 | | | | | -1.13 | |
| Apr-79 | | -1.00 | | | | |
| May-79 | | -0.63 | | | | |
| Jun-79 | | -2.10 | | | | |
| Jul-79 | | -1.78 | | | | |
| Aug-79 | | -0.50 | | | | |
| Sep-79 | | -1.07 | | | | |
| Nov-80 | -0.79 | | | | | |
| Dec-80 | -1.04 | | | | | |
| Jan-81 | -0.91 | | | | | |
| Feb-81 | -1.14 | | | | | |
| May-88 | | -1.13 | | | | |
| Jun-88 | | -1.50 | | | | |
| Jul-88 | | -1.43 | | | | |
| Jul-89 | | | | | -1.26 | |
| Aug-89 | | | | | -1.02 | |
| Sep-89 | | | | | -1.35 | |
| Oct-89 | | | | | -1.16 | |
| Nov-89 | | | | | -1.09 | |
| Dec-89 | | | | | -0.97 | |
| Jun-90 | | | | | -0.91 | |
| Jul-90 | | | | | -1.01 | |
| Aug-90 | | | | | -1.42 | |
| May-94 | | -1.05 | | | | |
| Jun-94 | | -1.76 | | | | |
| Jul-94 | | -1.30 | | | -1.27 | |
| Aug-94 | | | | | -1.10 | |
| Sep-94 | | | | | -1.12 | |

| M-Y | SPI | SPEI | SMRI | SRI | SSI | HADI |
|---|---|---|---|---|---|---|
| Jun-95 | | -1.60 | | | | |
| Jul-95 | | -2.06 | | | | |
| Aug-95 | | -1.09 | | | | |
| May-97 | | -0.96 | | | | |
| Jun-97 | | -1.21 | | | | |
| Jul-97 | | -1.41 | | | | |
| May-98 | | -1.44 | | | | |
| Jun-98 | | -0.92 | | | | |
| Jul-98 | | -1.27 | | | | |
| Aug-98 | | -1.10 | | | | |
| Oct-99 | | | -1.59 | | | |
| Nov-99 | | | -1.55 | | | |
| Dec-99 | | | -0.91 | | | |
| May-00 | | -1.04 | | | | |
| Jun-00 | | -1.15 | | | | |
| Jul-00 | | -1.11 | | | | |
| Jan-02 | -1.03 | | | -0.73 | | |
| Feb-02 | -1.63 | | -1.30 | -1.04 | | |
| Mar-02 | -0.40 | | -0.35 | -0.22 | | |
| Apr-02 | -0.73 | | -0.64 | -0.56 | | -0.71 |
| May-02 | -1.41 | | -1.25 | -1.32 | -1.19 | -1.01 |
| Jun-02 | -1.83 | | -1.61 | -1.84 | -2.29 | -1.06 |
| Jul-02 | | | | -1.06 | -2.31 | |
| Aug-02 | | | | -0.90 | -2.27 | |
| Sep-02 | | | | | -0.96 | |
| Jul-03 | | | | | -1.45 | |
| Aug-03 | | | | | -1.24 | |
| Sep-03 | | | | | -0.99 | |
| Oct-03 | | | | | -1.02 | |
| May-05 | | -1.14 | | | | |
| Jun-05 | | -1.38 | | | | |
| Jul-05 | | -2.17 | | | | |
| Aug-05 | | | | | | |
| Sep-05 | | | | | | |
| Mar-12 | -1.20 | | -0.76 | | | |
| Apr-12 | -0.53 | | -0.43 | | | |
| May-12 | -1.26 | | -1.09 | -1.01 | | |
| Jun-12 | -1.63 | | -1.40 | -1.57 | -1.59 | |
| Jul-12 | | | | -0.03 | -1.29 | |
| Aug-12 | | | | -1.19 | -2.02 | |
| Sep-12 | | | | -0.97 | -2.10 | |
| Oct-12 | | | | -1.08 | -1.73 | |
| Nov-12 | | | | -1.17 | -1.48 | |

**Figure 4.** Coincidence of dry years based on SPI, SPEI, SMRI, SRI, SSI, and HADI. Note: The dark and light brown colors show exceptional to abnormal droughts, respectively. The gray color shows normal condition.

Spatial distributions of droughts across the UCRB were also estimated and mapped using the monthly average from 2004 to 2014 (Figures 5–7). This period was selected since it was used for testing the prediction model. The driest months were June and November, based on the SPI. The SRI results showed the same months as the driest months through a year, plus December, in which droughts cover a significant portion of the UCRB (Northern and central parts). SSI and SPEI showed a more consistent pattern. Based on the SSI results, January is in wet conditions until May. Then, the drought starts covering the basin from the southern part and the entire UCRB in June. July to October remains in the severe dry conditions until November, when a change of status to milder drought conditions initiates from the southern parts. Then, the drought evidence started to disappear from the southern part of the UCRB. Based on the SPEI results (Figure 6), December and January are the months with the most extreme wet conditions. The wet condition starts to become milder in February, and abnormally dry conditions appeared in March. The most severe dry condition, which covers the entire UCRB, is observed in June. July and August remain dry. The evidence of abnormally wet conditions initiate in September from south of the UCRB. October, November, and December are moderately wet, severe to extreme wet, and exceptionally wet, respectively. Considering the SMRI results (Figure 6), the driest months are November, December, and June. The wettest months are July, August, and September. According to the HADI results (Figure 7), the entire UCRB is never in totally dry or wet conditions with the same severity. There is a mixture of different categories in each month. For instance, the basin is covered with a mix of normal and abnormally dry and moderate drought in January. The wet condition appeared in the southern, central, and northern parts of the UCRB in February, March, and April to June, respectively. The entire UCRB is covered by a normal condition except the southeastern area, which is in abnormally wet to moderate wet conditions in July. In August, the northern part is in abnormally wet condition, and the southern part is in abnormal to moderate droughts. A major portion of the UCRB is in normal condition in September and October except in the southern part and some other scattered areas. Normal condition covers the whole UCRB in November. Southern UCRB is abnormal to moderate wet conditions, and Central East and Northeastern are abnormal to moderate dry conditions in December.

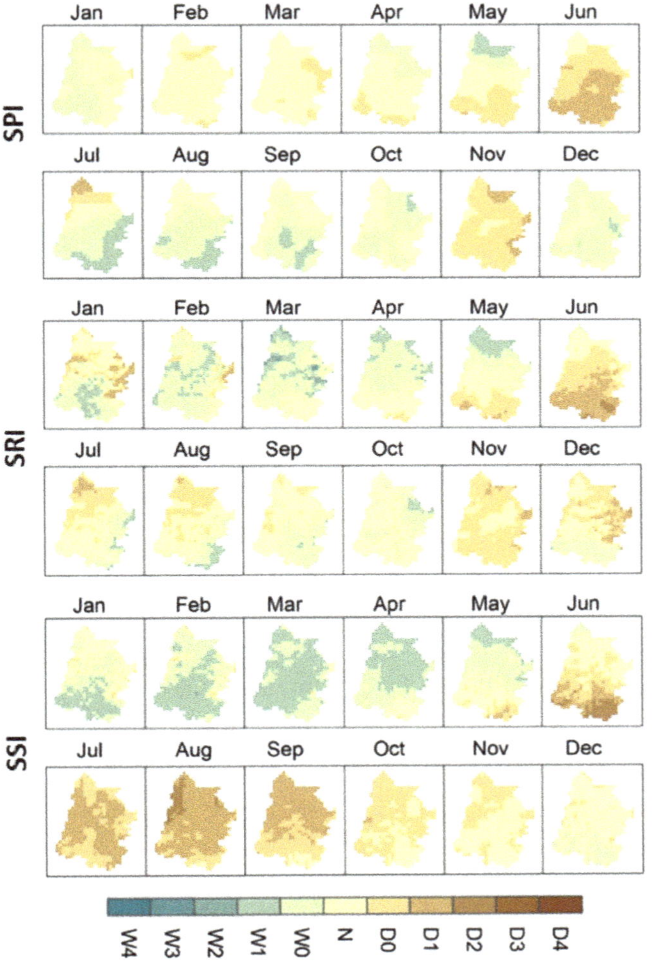


**Figure 5.** Spatial pattern of drought based on standardized univariate indices (SPI, SRI, and SSI) in the UCRB.

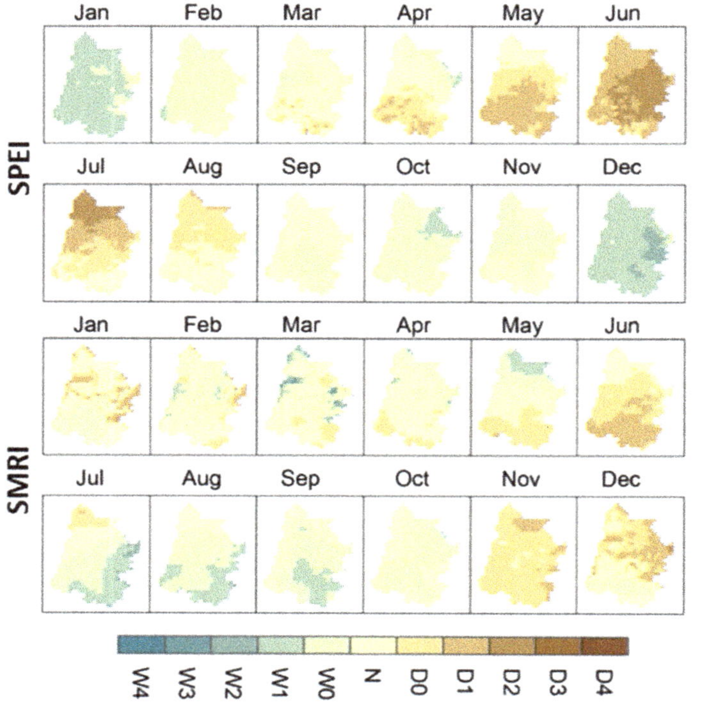


**Figure 6.** Spatial pattern of drought based on standardized bivariate indices (SPEI and SMRI) in the UCRB.

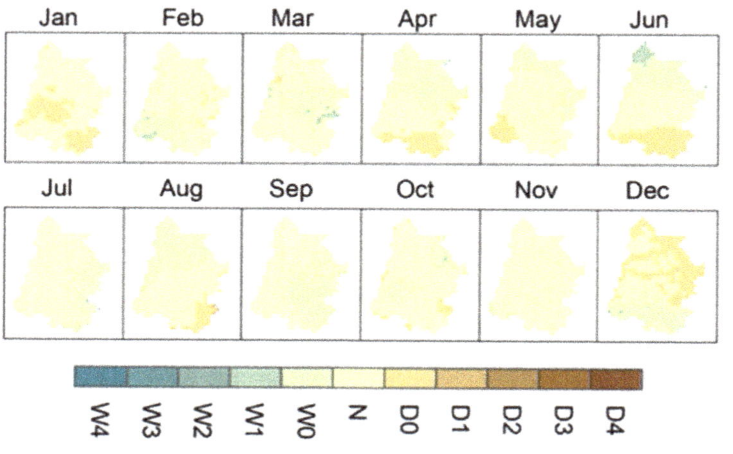


**Figure 7.** Spatial pattern of drought based on HADI in the UCRB.

Three major change points were detected in the temperature time series of the UCRB (Figure 8). The early years of the 1950s decade (March and August 1951, November 1952, and February 1954) were the primary time points for the first change points. The end of the 1970s (September 1978 and December 1979) and the beginning of the 1980s (May and October 1980) were detected as the second change points. The third change points were mainly identified at the end of the 1990s (September 1998 and December 1999) and the beginning of the 21st century (May and October 2000 and September 2003). Both the monthly temperature time series and the 12-month moving average show some changes in the statistical properties. Based on these three detected change points, the average temperature values in the four timeslots are 5.88, 6.32, 6.71, and 6.63. The corresponding standard deviations are 8.84, 8.97, 8.8, and 9.04.

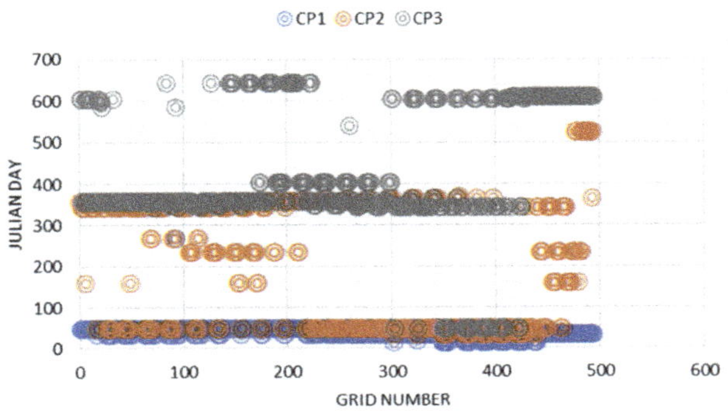


| Change Point 1 (CP1) | | Change Point 2 (CP2) | | Change Point 3 (CP3) | |
|---|---|---|---|---|---|
| Julian day | Month-Year | Julian day | Month-Year | Julian day | Month-Year |
| **15** | Mar-51 | **35** | Nov-52 | **50** | Feb-54 |
| **20** | Aug-51 | **50** | Feb-54 | **345** | Sep-78 |
| **35** | Nov-52 | **160** | Apr-63 | **350** | Feb-79 |
| **50** | Feb-54 | **235** | Jul-69 | **360** | Dec-79 |
| **270** | Jun-72 | **270** | Jun-72 | **365** | May-80 |
| | | **345** | Sep-78 | **405** | Sep-83 |
| | | **360** | Dec-79 | **540** | Dec-94 |
| | | **365** | May-80 | **585** | Sep-98 |
| | | **370** | Oct-80 | **600** | Dec-99 |
| | | **525** | Sep-93 | **605** | May-00 |
| | | | | **610** | Oct-00 |
| | | | | **645** | Sep-03 |

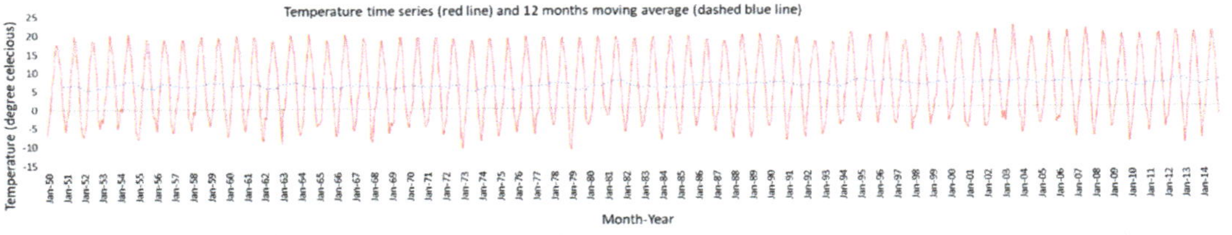


**Figure 8.** The first, second, and third change points of different grids, temperature time series and the 12-month moving average in the UCRB.

### 3.2. Drought Prediction Results

Figure 9 shows the results for the comparison of the performances of SVR and ES-CBS-SVR in predicting drought in the UCRB. The violin plot in Figure 9a shows that $R^2$ values are quite similar and close to 1 in both models. However, the $R^2$ for ES-CBS-SVR is more accumulated around 1, demonstrating the best performance. The horizontal bar in the same figure also shows that ES-CBS-SVR performs better than SVR in a higher percentage of the UCRB. MSE and RMSE also showed the same results. However, the performance of SVR is better than ES-CBS-SVR in a slightly higher portion of the UCRB based on the MAE. The SPEI results show the same pattern for both models based on all the performance indicators. This can be interpreted as the similarities (not significant differences) of the performance indicators' values. However, the horizontal bar shows a significantly higher percentage of the UCRB with better performance of the ES-CBS-SVR. Although the differences between performance indicators were not significant according to the violin plots, the differences in the percentage of the area in the performance of the prediction models were substantial. The performance of SVR was better than ES-CBS-SVR for SMRI and SRI. When SSI was used in the prediction models, the ES-CBS-SVR was better than the SVR for all the performance indicators except the MAE. Like SPEI, when the HADI was employed in the prediction models, the performance of the ES-CBS-SVR was better in a significantly higher portion of the UCRB. Therefore, the performance of the ES-CBS-SVR was better in a higher percentage of the UCRB when SPI, SPEI, SSI, and HADI were used.

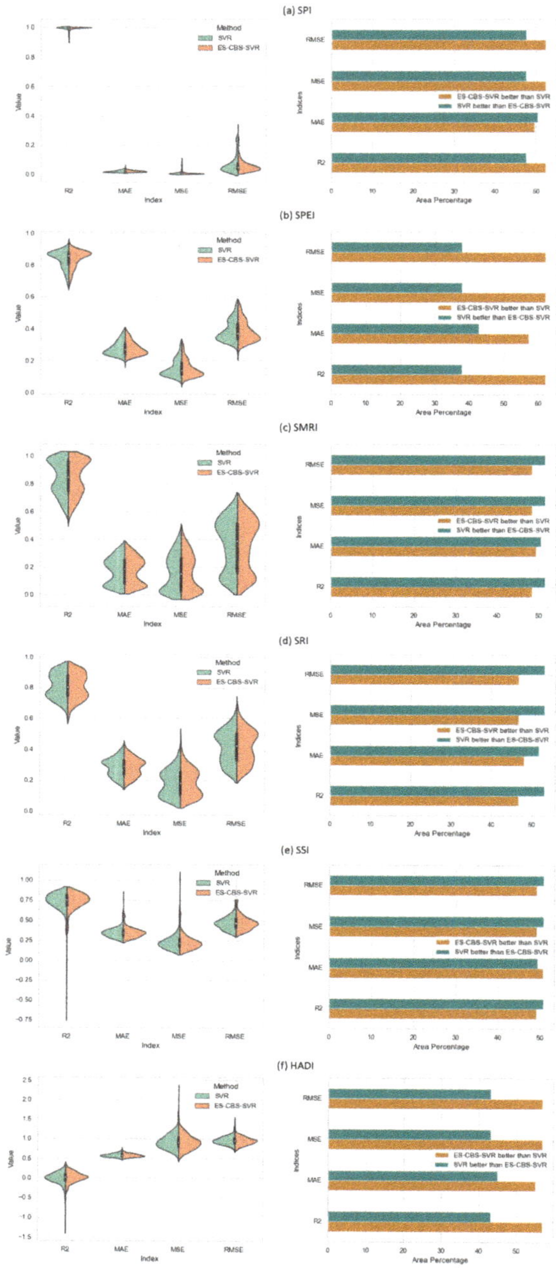


**Figure 9.** Comparison of the performances of SVR and ES-CBS-SVR for (**a**) SPI, (**b**) SPEI, (**c**) SMRI, (**d**) SRI, (**e**) SSI, and (**f**) HADI. The graph on the left side for each drought index is a violin plot, while the horizontal bar graph represents a comparison of the number of grids in the study area with accurate predictions for each drought index.

## 4. Discussion

The identified droughts based on the SPI, SPEI, SMRI, SRI, SSI, and HADI are similar to the other drought studies [37,45]. However, unlike Timilsena and Piechota (2008) [46], which demonstrated that the average drought duration in the UCRB was 3–14 years, the most prolonged dry periods lasted less, considering the wet conditions among the dry periods. The SPEI results show that droughts are more prevalent in the warm seasons (May, June, and July) in the UCRB. Droughts are aggravated by enhanced, globally warmed temperatures that reduce spring snowpack and late spring and summer soil moisture [20]. The critical role of temperature in the SPEI and thus the impact of temperature on evapotranspiration and water deficit is associated with the common droughts in the warm seasons.

Drought from June to August 2002 is the longest consistent exceptional drought based on SSI. SSI also shows that a two-month exceptional drought occurred between August and September 2012. However, the other drought indices show less consistency for exceptional drought. Since SMRI is related to snowmelt and SPI did not identify any drought, the droughts in 1976, 2002, and 2012 can be classified as snow droughts. Lack of snow or late snowmelt due to later than normal increases in temperature can cause a deficit in water originating from snowmelt.

The results for HADI showed that the most extended dry period was from April to June 2002. Remarkably, it coincides with the sole period in the study period (1950 to 2014) that SPI, SMRI, SRI, and SSI showed extreme drought conditions. The results of Woodhouse et al. (2006) [45] also indicated that this dry period was long and severe. They showed evidence of drought disappearance from the UCRB in the winter of 2004–2005. However, the drought impacts (e.g., abnormally low water level in the major reservoirs on the Colorado River) remained until 2005 [45]. HADI is estimated based on anomalies in rainfall, snowmelt, runoff, and soil moisture, and this coincidence verifies the accuracy of the HADI and emphasizes its application case. Thus, HADI can be a good substitute for a combined application of SPI, SMRI, SRI, and SSI.

The results demonstrated a spatial pattern in drought appearance and disappearance. The southern part of the UCRB is the first area affected by drought, and the disappearance of drought is also initiated from this area. Woodhouse et al. (2006) [45] also indicated that the southern portion of the UCRB gained a measure of drought relief in the winter of 2004–2005. The less evident spatial pattern and a mixture of drought categories in the UCRB are associated with the nature of the HADI, which reflects the anomalies in different hydroclimatic variables.

SMRI reflects the anomalies in immediately available water, which is the sum of rainfall and snowmelt. The dry conditions in the warm seasons (e.g., June) can be associated with a rainfall deficit or high temperature. Since June is one of the driest months based on the SPEI, the relatively high temperature in June can be one of the primary drivers. In contrast, the dry conditions in the cold seasons (November) can be related to the precipitation types in the form of snowfall or, generally, precipitation deficit.

The changes in the statistical properties of the temperature time series confirm the accuracy of the detected change points and the existence of non-stationarity. According to the climatic change scenarios, a shift in the ratio of snowfall to rainfall and earlier snowmelt and runoff [47,48] will likely compound the strain on water resources throughout the entire Colorado River Basin [45]. Thus, it is crucial to overcome this non-stationarity in drought analyses.

The significantly higher percentage of the UCRB with better performance of the ES-CBS-SVR than SVR using SPEI is associated with the importance of temperature in this index. The ES-CBS-SVR is developed based on the variability/changes in temperature time series. Thus, the performance of the ES-CBS-SVR model was better when the model used SPEI. Unlike other indices, HADI was the same as SPEI. HADI has been accustomed to cold climate regions, and temperature also plays a vital role in this index.

## 5. Conclusions

Six drought indices, SPI, SPEI, SMRI, SRI, SSI, and HADI, were used to identify droughts in the UCRB. A customized drought categorization was used to consider both spatial and temporal distributions of drought in drought categorization. The non-stationarity issue in a warming climate in drought prediction was removed by developing ES-CBS-SVR. The foundation of ES-CBS-SVR is based on dividing the non-stationary dataset into stationary subsets based on the change points in the temperature time series. The ES-CBS-SVR uses category-based scoring for tuning the hyperparameters.

Droughts are more prevalent in the warm seasons (May, June, and July) in the UCRB based on the SPEI due to the critical role of temperature in SPEI. The dry conditions in June or other months in the warm seasons based on the SMRI can be associated with the deficit of rainfall or high temperature. Other studies also confirmed that increasing temperature in the warming climate plays a prominent role in recent droughts [49]. SSI identified the longest consistent, exceptional drought. SPI, SPEI, SMRI, SRI, and HADI show relatively less consistency for exceptional drought. The SMRI identified droughts in 1976, 2002, and 2012 that can be classified as snow droughts. HADI identified the most prolonged dry period from April to June 2002. SPI, SMRI, SRI, and SSI showed extreme drought conditions in the same period. Thus, HADI can be a good substitute for a combined application of SPI, SMRI, SRI, and SSI, which reflect anomalies in precipitation, rainfall and snowmelt, runoff, and soil moisture, respectively. A spatial pattern in drought appearance and disappearance was identified in the UCRB, demonstrating the southern part of the UCRB is the initial point of drought.

The existence of non-stationarity in the UCRB was identified in the early years of the 1950s, the end of the 1970s, the beginning of the 1980s, the end of the 1990s, and the beginning of the 21st century. The changes in the statistical properties of the temperature time series confirm the accuracy of the detected change points.

The performance of the ES-CBS-SVR was better in a higher portion of the UCRB when SPI, SPEI, SSI, and HADI were used. The significantly higher percentage of the UCRB with better performance of the ES-CBS-SVR than SVR using SPEI and HADI is associated with the importance of temperature in these indices.

This study covers different types of droughts and six drought indices: SPI, SPEI, SMRI, SRI, SSI, and HADI. SVR was used for drought prediction, as this model was highly suggested in other studies. Thus, the application of other drought indices and other prediction models, such as artificial neural networks and deep learning, can be considered for future studies. Finally, the current study focused primarily on droughts in the UCRB. To generalize the applicability of the findings, future studies should expand the analysis to other basins with diverse climatic and hydrologic conditions. This would help to test the adaptability and performance of ES-CBS-SVR and the indices used in this study under varying environmental and anthropogenic influences.

**Author Contributions:** Conceptualization, M.H.B. and N.Z.; methodology, M.H.B. and N.Z.; software, M.H.B.; validation, M.H.B., H.H., T.A., S.P. and N.Z.; formal analysis, M.H.B., H.H., T.A., S.P. and N.Z.; investigation, M.H.B., H.H., T.A., S.P. and N.Z.; resources, M.H.B., H.H., T.A., S.P. and N.Z.; data curation, M.H.B. and H.H.; writing—original draft preparation, M.H.B.; writing—review and editing, M.H.B., H.H., T.A., S.P., N.Z. and J.J.; visualization, M.H.B.; supervision, J.J.; project administration, J.J.; funding acquisition, J.J. All authors have read and agreed to the published version of the manuscript.

**Funding:** This project was supported by the Bureau of Land Management (Award #: L17AC00125).

**Institutional Review Board Statement:** Not applicable.

**Informed Consent Statement:** Not applicable.

**Data Availability Statement:** The datasets for this research are available at the GLDAS Noah Land Surface Model L4 3 hourly 0.25 × 0.25 degree V2.1: https://disc.gsfc.nasa.gov/datasets?keywords=

GLDAS (26 August 2022). The modeling data generated from this study will be available for any interested readers.

**Conflicts of Interest:** The authors declare no conflicts of interest.

## References


1. Castle, S.L.; Thomas, B.F.; Reager, J.T.; Rodell, M.; Swenson, S.C.; Famiglietti, J.S. Groundwater depletion during drought threatens future water security of the Colorado River Basin. *Geophys. Res. Lett.* **2014**, *41*, 5904–5911. [CrossRef]
2. Barnett, T.P.; Pierce, D.W. Sustainable water deliveries from the Colorado River in a changing climate. *Proc. Natl. Acad. Sci. USA* **2009**, *106*, 7334–7338. [CrossRef]
3. Tillman Fred, D.; Day, N.K.; Miller, M.P.; Miller, O.L.; Rumsey, C.A.; Wise, D.R.; Longley, P.C.; McDonnell, M.C. A Review of Current Capabilities and Science Gaps in Water Supply Data, Modeling, and Trends for Water Availability Assessments in the Upper Colorado River Basin. *Water* **2022**, *14*, 3813. [CrossRef]
4. Woodhouse, C.A.; Smith, R.M.; McAfee, S.A.; Pederson, G.T.; McCabe, G.J.; Miller, W.P.; Csank, A. Upper Colorado River Basin 20th century droughts under 21st century warming: Plausible scenarios for the future. *Clim. Serv.* **2021**, *21*, 100206. [CrossRef]
5. Gray, S.T.; Jeffrey, J.L.; Woodhouse, C.A. Millennial-Length Records of Streamflow From Three Major Upper Colorado River Tributaries. *J. Am. Water Resour. Assoc. (JAWRA)* **2011**, *47*, 702–712. [CrossRef]
6. Reynolds, L.V.; Shafroth, P.B.; LeRoy Poff, N. Modeled intermittency risk for small streams in the upper Colorado River basin under climate change. *J. Hydrol.* **2015**, *523*, 768–780. [CrossRef]
7. Harding, B.L.; Sangoyomi, T.B.; Payton, E.A. Impacts of a severe sustained drought on Colorado Rwer Water Resources. *J. Am. Water Resour. Assoc.* **1995**, *31*, 815–824. [CrossRef]
8. Meko, D.M.; Woodhouse, C.A. Tree-ring footprint of joint hydrologic drought in Sacramento and Upper Colorado River basins, Western USA. *J. Hydrol.* **2005**, *308*, 196–213. [CrossRef]
9. Meko, D.M.; Woodhouse, C.A.; Baisan, C.A.; Knight, T.; Lukas, J.J.; Hughes, M.K.; Salzer, M.W. Medieval drought in the upper Colorado River Basin. *Geophys. Res. Lett.* **2007**, *34*, L10705. [CrossRef]
10. Timilsena, J.; Piechota, T.C.; Hidalgo, H.; Tootle, G. Five Hundred Years of hydrological drought in the upper Colorado River Basin. *J. Am. Water Resour. Assoc.* **2007**, *43*, 798–812. [CrossRef]
11. Salehabadi, H.; Tarboton, D.G.; Udall, B.; Wheeler, K.G.; Schmidt, J.C. An assessment of potential severe droughts in the Colorado River Basin. *JAWRA J. Am. Water Resour. Assoc.* **2022**, *58*, 1053–1075. [CrossRef]
12. McCabe, G.J.; Wolock, D.M.; Woodhouse, C.A.; Pederson, G.T.; McAfee, S.A.; Gray, S.; Csank, A. Basinwide Hydroclimatic Drought in the Colorado River Basin. *Earth Interact.* **2020**, *24*, 1–20. [CrossRef]
13. Miller, W.P.; Piechota, T.C. Trends in Western U.S. Snowpack and Related Upper Colorado River Basin Streamflow1. *JAWRA J. Am. Water Resour. Assoc.* **2011**, *47*, 1197–1210. [CrossRef]
14. Vicente-Serrano, S.M.; Beguería, S.; López-Moreno, J.I. A multiscalar drought index sensitive to global warming: The standardized precipitation evapotranspiration index. *J. Clim.* **2010**, *23*, 1696–1718. [CrossRef]
15. McKee, T.B.; Doesken, N.J.; Kleist, J. The relationship of drought frequency and duration to time scales. In Proceedings of the 8th Conference on Applied Climatology, American Meteorological Society, Anaheim, CA, USA, 17–22 January 1993.
16. Abatan, A.A.; Gutowski, W.J.; Ammann, C.M.; Kaatz, L.; Brown, B.G.; Buja, L.; Bullock, R.; Fowler, T.; Gilleland, E.; Gotway, J.H. Multiyear droughts and pluvials over the upper Colorado River Basin and associated circulations. *J. Hydrometeorol.* **2017**, *18*, 799–818. [CrossRef]
17. Woodhouse, C.A.; Pederson, G.T.; Morino, K.; McAfee, S.A.; McCabe, G.J. Increasing influence of air temperature on upper ColoradoRiver streamflow. *Geophys. Res. Lett.* **2016**, *43*, 2174–2181. [CrossRef]
18. Udall, B.; Overpeck, J. Thetwenty-first century Colorado River hot drought and implications for the future. *Water Resour. Res.* **2017**, *53*, 2404–2418. [CrossRef]
19. McCabe, G.J.; Wolock, D.M.; Pederson, G.T.; Woodhouse, C.A.; McAfee, S. Evidence that recent warming is reducing Upper Colorado River flows. *Earth Interact.* **2017**, *21*, 1–14. [CrossRef]
20. Cayan, D.R.; Das, T.; Pierce, D.W.; Barnett, T.P.; Tyree, M.; Gershunov, A. Future dryness in the southwest US and the hydrology of the early 21st Century drought. *Proc. Natl. Acad. Sci. USA* **2010**, *107*, 21271–21276. [CrossRef]
21. Tang, C.; Piechota, T.C. Spatial and temporal soil moisture and drought variability in the upper Colorado River Basin. *J. Hydrol.* **2009**, *379*, 122–135. [CrossRef]
22. Mohsin, M.; Pilz, J. Stochastic model for drought analysis of the Colorado River Basin. *Stoch Environ. Res Risk Assess* **2021**, *35*, 1637–1648. [CrossRef]
23. Madadgar, S.; Moradkhani, H. A Bayesian Framework for Probabilistic Seasonal Drought Forecasting. *J. Hydrometeorol.* **2013**, *14*, 1685–1705. [CrossRef]
24. Modarres, R. Streamflow drought time series forecasting. *Stoch. Environ. Res. Risk Assess.* **2007**, *21*, 223–233. [CrossRef]
25. Ellis, A.W.; Goodrich, G.B.; Garfin, G.M. A hydroclimatic index for examining patterns of drought in the Colorado River Basin. *Int. J. Climatol.* **2009**, *30*, 236–255. [CrossRef]
26. Bazrkar, M.H.; Chu, X. Ensemble Stationary-based Support Vector Regression for Drought Prediction under Changing Climate. *J. Hydrol.* **2021**, *603*, 127059. [CrossRef]

27. Murphy, K.W.; Ellis, A.W. An assessment of the stationarity of climate and stream flow in watersheds of the Colorado River Basin. *J. Hydrol.* **2014**, *509*, 454–473. [CrossRef]
28. Brunner, M.I.; Slater, L.; Tallaksen, L.M.; Clark, M. Challenges in modeling and predicting floods and droughts: A review. *WIREs Water* **2021**, *8*, e1520. [CrossRef]
29. Hao, Z.; Singh, V.P.; Xia, Y. Seasonal Drought Prediction: Advances, Challenges, and Future Prospects. *Rev. Geophys.* **2018**, *56*, 108–141. [CrossRef]
30. Ju, X.; Wang, Y.; Wang, D.; Singh, V.P.; Xu, P.; Wu, J.; Ma, T.; Liu, J.; Zhang, J. A time varying drought identification and frequency analyzation method: A case study of Jinsha River Basin. *J. Hydrol.* **2021**, *603*, 126864. [CrossRef]
31. Mukherjee, S.; Mishra, A.; Trenberth, K.E. Climate Change and Drought: A Perspective on Drought Indices. *Curr. Clim. Chang. Rep.* **2018**, *4*, 145–163. [CrossRef]
32. Talsma, C.J.; Bennett, K.E.; Vesselinov, V.V. Characterizing drought behavior in the Colorado River Basin using unsupervised machine learning. *Earth Space Sci.* **2022**, *9*, e2021EA002086. [CrossRef]
33. McCabe, G.J.; Betancourt, J.L.; Hidalgo, H.G. Associations of decadal to Multidecadal Sea-surface temperature variability with upper Colorado River flow1. *JAWRA J. Am. Water Resour. Assoc.* **2007**, *43*, 183–192. [CrossRef]
34. Adarsh, S.; Adarsh, S.; Reddy, M.J. Multiscale characterization and prediction of monsoon rainfall in India using Hilbert–Huang transform and time-dependent intrinsic correlation analysis. *Meteorol. Atmos. Phys.* **2018**, *130*, 667–688. [CrossRef]
35. Achour, K.; Meddi, M.; Zeroual, A.; Bouabdelli, S.; Maccioni, P.; Moramarco, T. Spatio-temporal analysis and forecasting of drought in the plains of northwestern Algeria using the standardized precipitation index. *J. Earth Syst. Sci.* **2020**, *129*, 42. [CrossRef]
36. Bazrkar, M.H.; Chu, X. Development of category-based scoring support vector regression (CBS-SVR) for drought prediction. *J. Hydroinformatics* **2022**, *24*, 202–222. [CrossRef]
37. Mix, K.; Groeger, A.W.; Lopes, V.L. Impacts of dam construction on streamflows during drought periods in the upper Colorado River Basin, Texas. *Lakes Reserv. Sci. Policy Manag. Sustain. Use* **2016**, *21*, 329–337. [CrossRef]
38. Woodhouse, C.A.; Pederson, G.T. Investigating runoff efficiency in upper Colorado river streamflow over past centuries. *Water Resour. Res.* **2018**, *54*, 286–300. [CrossRef]
39. Truong, C.; Oudre, L.; Vayatis, N. Selective review of offline change point detection methods. *Signal Process.* **2020**, *167*, 107299. [CrossRef]
40. Vapnik, V. *The Nature of Statistical Learning Theory*; Springer: New York, NY, USA, 1995.
41. Bazrkar, M.H.; Zhang, J.; Chu, X. Hydroclimatic aggregate drought index (HADI): A new approach for identification and categorization of drought in cold climate regions. *Stoch. Environ. Res. Risk Assess.* **2020**, *34*, 1847–1870. [CrossRef]
42. Farahmand, A.; AghaKouchak, A. A generalized framework for deriving nonparametric standardized drought indicators. *Adv. Water Resour.* **2015**, *76*, 140–145. [CrossRef]
43. Svoboda, M.; LeComte, D.; Hayes, M.; Heim, R.; Gleason, K.; Angel, J.; Stooksbury, D. The drought monitor. *Bull. Am. Meteorol. Soc.* **2002**, *83*, 1181–1190. [CrossRef]
44. Rodell, M.; Houser, P.R.; Jambor, U.; Gottschalck, J.; Mitchell, K.; Meng, C.-J.; Arsenault, K.; Cosgrove, B.; Radakovich, J.; Bosilovich, M.; et al. The Global Land Data Assimilation System. *Bull. Amer. Meteor. Soc.* **2004**, *85*, 381–394. [CrossRef]
45. Woodhouse, C.A.; Gray, S.T.; Meko, D.M. Updated streamflow reconstructions for the Upper Colorado River Basin. *Water Resour. Res.* **2006**, *42*, 16. [CrossRef]
46. Timilsena, J.; Piechota, T. Regionalization and reconstruction of snow water equivalent in the upper Colorado River Basin. *J. Hydrol.* **2008**, *352*, 94–106. [CrossRef]
47. Cayan, D.R.; Kammerdiener, S.A.; Dettinger, M.D.; Caprio, J.M.; Peterson, D.H. Changes in the onset of spring in the western United States. *Bull. Am. Meteorol. Soc.* **2001**, *82*, 399–415. [CrossRef]
48. Stewart, I.T.; Cayan, D.R.; Dettinger, M.D. Changes in snowmelt runoff timing in western North America under a "business as usual" climate change scenario. *Clim. Chang.* **2004**, *62*, 217–232. [CrossRef]
49. Xiao, M.; Udall, B.; Lettenmaier, D.P. On the causes of declining Colorado River streamflows. *Water Resour. Res.* **2018**, *54*, 6739–6756. [CrossRef]

atmosphere

MDPI

*Article*

# A Snow-Based Hydroclimatic Aggregate Drought Index for Snow Drought Identification

**Mohammad Hadi Bazrkar [1], Negin Zamani [2] and Xuefeng Chu [3,*]**

[1] Blackland Research and Extension Center, Texas A&M AgriLife Research, Temple, TX 76502, USA; hadi.bazrkar@ag.tamu.edu
[2] Department of Civil and Architectural Engineering, Texas A&M University Kingsville, Kingsville, TX 78363, USA; negin.zamani@students.tamuk.edu
[3] Department of Civil and Environmental Engineering, North Dakota State University, P.O. Box 6050, Fargo, ND 58108, USA
* Correspondence: xuefeng.chu@ndsu.edu

**Abstract:** Climate change has increased the risk of snow drought, which is associated with a deficit in snowfall and snowpack. The objectives of this research are to improve drought identification in a warming climate by developing a new snow-based hydroclimatic aggregate drought index (SHADI) and to assess the impacts of snowpack and snowmelt in drought analyses. To derive the SHADI, an R-mode principal component analysis is performed on precipitation, snowpack, surface runoff, and soil water storage. Then, a joint probability distribution function of drought frequencies and drought classes, conditional expectation, and *k*-means clustering are used to categorize droughts. The SHADI was applied to the Red River of the North Basin (RRB), a typical cold climate region, to characterize droughts in a mostly dry period from 2003 to 2007. The SHADI was compared with the hydroclimatic aggregate drought index (HADI) and U.S. drought monitor (USDM) data. Cluster analysis was also utilized as a benchmark to compare the results of the HADI and SHADI. The SHADI showed better alignment with cluster analysis results than the HADI, closely matching the identified dry/wet conditions in the RRB. The major differences between the SHADI and HADI were observed in cold seasons and in transition periods (dry to wet or wet to dry). The derived variable threshold levels for different categories of drought based on the SHADI were close to, but different from, those of the HADI. The SHADI can be used for short-term lead prediction of droughts in cold climate regions and, in particular, can provide an early warning for drought in the warming climate.

**Keywords:** snow drought; drought; cold climate; drought monitoring; Red River of the North Basin; snow-based hydroclimatic aggregate drought index (SHADI)

## 1. Introduction

Climate change and global warming have caused a shrinkage in average snow cover and montane snowpack [1–3]. The function of snowpack is similar to that of a seasonal natural reservoir, where snowmelt leads the equivalent stored water to be released by an increase in temperature. Shifts in temperature and precipitation patterns due to climate change have heightened the risk of snow drought [4,5].

Snow drought and its impact on water resources availability were introduced by Schneider & Matson (1977) [6] and Wiesnet (1981) [7]. Snow drought occurs in a period characterized by unusually low snowpack (depth of accumulated snow) resulting from either below-average cold-season precipitation or insufficient snow accumulation despite near-normal precipitation levels, typically due to warm temperatures causing rain instead of snow or early snowmelt [8]. Based on the definition of snow drought, the drought drivers include (1) quantity and (2) timing. The first drought driver is associated with the deficit in the amount of precipitation and/or high or low levels of temperatures, while the second drought driver is characterized by an early or late occurrence of hydroclimatic processes such as snowfall and snowmelt. Depending on the primary driver of drought—temperature or precipitation—snow drought can be categorized into two types: temperature-driven or precipitation-driven. Temperature-driven drought occurs when warmer-than-normal temperatures reduce snow accumulation, even if precipitation levels are near or above average. In this case, precipitation falls as rain instead of snow, or accumulated snow melts prematurely. Precipitation-driven drought occurs when there is below-normal precipitation during the snow accumulation season, leading to reduced snowpack regardless of temperature conditions [9–11]. Harpold et al. (2017) [12] also classified snow drought into dry snow drought and warm snow drought based on these climatic drivers. Dry snow droughts are caused by a deficit in winter precipitation, while above-normal winter temperatures (consequently, late onset of snow season), midseason snowmelt or rainfall events, and early spring snowmelt cause warm snow drought [13]. For instance, a temperature threshold of −3.1 °C and 1.4 °C led to a rapid increase in the risk of warm snow drought in southwestern Canada [14]. Therefore, the first step in drought identification is to understand the dominant hydroclimatic processes and drought drivers and types [15]. Carrying out this essential step ensures the use of the correct tools or drought indices [16].

Each region has its own distinct drought drivers [17]. For example, low precipitation, high temperature, and low initial soil moisture caused a considerable reduction in snowmelt runoff in the Canadian prairies [18]. Generally, a low amount of precipitation (snowfall and rainfall) because of high temperature in winter may bring about "warm snow season drought". Moreover, in regions where the winter temperatures normally are far below zero for a long time (e.g., more than 6 months of a year), a mild anomaly in winter temperature may affect snow accumulation, snow water equivalent, and streamflow. In such areas, a lack of snow accumulation due to snowfall deficit or high temperature in winter provokes a shift in the timing of snowmelt and causes "snowmelt drought" [19,20]. These drought classes accentuate the necessity of consideration of snow storage and release (snowmelt) in identification of drought in regions with snow accumulation and snowmelt processes.

Snow drought can cause some challenges for water resource management in cold climate regions. On the one hand, the lack of snowpack causes a reduction in available water resources in warm seasons, and on the other hand, a substitution of rain for snow in unusually warm winters leads to a higher risk of flood in mountainous regions. Correspondingly, snow drought can cause serious ecological and socioeconomic impacts [14]. For example, the agricultural losses caused by the 2015 dry snow drought in the Sierra Nevada region of the U.S. were estimated at approximately USD 1.84 billion, resulting from a 20–60% reduction in normal precipitation [21]. In this region, extreme early season precipitation, frequent rain-on-snow events, and low-precipitation years were the main causes of snow drought [22]. Therefore, understanding the mechanisms of droughts and their drivers in cold climate regions is important to identify snow droughts and to mitigate their impacts.

To mitigate the impacts of droughts in the changing climate, a special drought index for the identification of snow droughts is essential. There exist a few drought indices which account for the anomalies in snowpack. Huning and Aghakouchak (2020) [23] developed a standardized snow water equivalent index (SWEI) by considering snow water equivalent (SWE) to assess global snow drought. Since a below-normal SWE in cold seasons can be compensated by a high antecedent soil water storage or an upcoming rainfall event in warm seasons, consideration of other hydroclimatic variables can provide more information about drought. The surface water supply index (SWSI) [24] and aggregate drought index (ADI) [25] account for snowpack amounts in addition to other hydroclimatic variables. In contrast, other indices account for available water instead of the stored water sources. For example, the standardized snowmelt and rainfall index (SMRI) [26] and hydroclimatic aggregate drought index (HADI) [27] consider snowmelt and rainfall. Some studies demonstrated the critical roles of snow-based drought indices [23] and their capabilities for short-term lead prediction of drought [28] in a changing climate. However, some researchers (e.g., Livneh and Badger 2020) [29] found drought less predictable by using snow-based approaches under declining future snowpack in a warming climate.

The key question is which drought index is most applicable in a snow-dominated, cold climate region. To address this, it is crucial to compare drought indices that consider stored water sources (e.g., snowfall and accumulated snow) with those that examine anomalies in available water resources (e.g., rainfall and snowmelt) for identification and short-term lead prediction of droughts. The objectives of this research are (1) to improve drought identification in cold climate regions and in the warming climate by developing a new snow-based hydroclimatic aggregate drought index (SHADI) and (2) to assess the impacts of snowpack and snowmelt in drought analyses by comparing the SHADI with the HADI. The structure of this paper is organized as follows: Section 1 introduces the background, objectives and scope of the study. Section 2 outlines the methodology and describes the study area. The results of the research are presented in Section 3, followed by the conclusion and recommendations in Section 4.

## 2. Methodology

### 2.1. Study Area

The Red River of the North Basin (RRB) (Figure 1), located in the Northern Great Plains, covers over 90,000 km$^2$ in the states of Minnesota, North Dakota (ND), and South Dakota. In the RRB, cold and dry winters are common due to a strong continental climate, a very low amount of air moisture, and cold intrusions from the polar and arctic regions [30]. Remarkably, the precipitation in the six cold months from October through March accounts for only about 22% of the annual precipitation. In contrast, a slow transition in low-level wind flow brings about a high amount of moisture from the Gulf of Mexico in spring. Therefore, springs and summers are generally warm and wet [30]. The climate divisions (CDs) in the RRB are shown in Figure 1. CDs 3203, 3206, and 2101 are entirely within the boundaries of the RRB. This is why these CDs were selected for analyzing the results. In CD 3203, the average, maximum, and minimum temperatures were recorded as 3.16 °C, 9.27 °C, and −2.88 °C, respectively, with a total precipitation of 480 mm. For CD 3206, these values were 4.44 °C, 10.61 °C, and −1.72 °C, along with 510 mm of precipitation. Similarly, in CD 2101, the average temperature was 3.55 °C, the maximum temperature reached 9.55 °C, the minimum dropped to −2.44 °C, and the precipitation totaled 559 mm [31]. In addition to climate factors, the CDs were organized based on geographic features, drainage basins, river districts, and/or designated forecast responsibility areas [32]. A divisional dataset has been assembled containing year-monthly averages (daily averages for each year and month) of temperature and water-equivalent precipitation for each division in the contiguous United States dating back to 1895. From this foundational dataset, statistics on drought, prolonged wet weather events (moisture anomalies), and heating/cooling degree days have been derived. The National Climatic Data Center (NCDC) uses this dataset in its monthly Climate Variations Bulletin [32].

According to the 118-year data (1901–2019), the mean discharge of the Red River is 12.77 m$^3$/s, and the minimum and maximum discharges at the Fargo Station are 0.045 m$^3$/s (1937, known as the Great Drought) and 226.8 m$^3$/s (1998), respectively [33]. The lowest precipitation and the highest temperature in the cold seasons occurred in the 1990s, which can be referred to as snow drought. The precipitation in the cold seasons in the last two decades was relatively low. Therefore, a drought index with the capability of addressing snow drought is required. To test the performance of the SHADI, it was applied to the RRB for a period from 2003 to 2007, the longest dry period after 2000 in the area [34].

### 2.2. Calculation of SHADI and Input Data

Precipitation, snowpack, surface runoff, and soil water storage of the root zone are used to derive the SHADI. Snowpack is the primary variable in the identification of snow drought. Thus, the SHADI can potentially address the anomalies in stored water as a snow drought index. The other hydroclimatic variables (i.e., precipitation, surface runoff, and soil water storage) are also integrated to address other types of droughts. Similar to the HADI [27], the SHADI accounts for the impacts of available water or stored water on drought characterization. However, the SHADI uses precipitation and snowpack, while the HADI uses rainfall and snowmelt. The daily snowpack, surface runoff, and soil water storage simulated by a grid-based hydrologic model (GHM) (Chu et al., 2019 [34]) are used after being converted to monthly values. The GHM model simulates key hydroclimatic processes in cold climate regions, including snow, snow accumulation, frozen soil, and snowmelt, at a spatial resolution of 4 km. The GHM models rainfall-runoff and snowfall-snowmelt processes based on a daily mean temperature threshold. In a snow-covered area, snowmelt is simulated using the degree-day method, where melting begins when the daily mean temperature exceeds a specified base temperature. The curve number (CN) is adjusted to account for frozen soil conditions. Moreover, as the RRB is a region dominated by surface depressions, the model includes a component of surface depression storage which is identified by a delineation algorithm [35]. Figure 2 presents the flowchart for calculating the HADI and SHADI and for categorizing droughts.

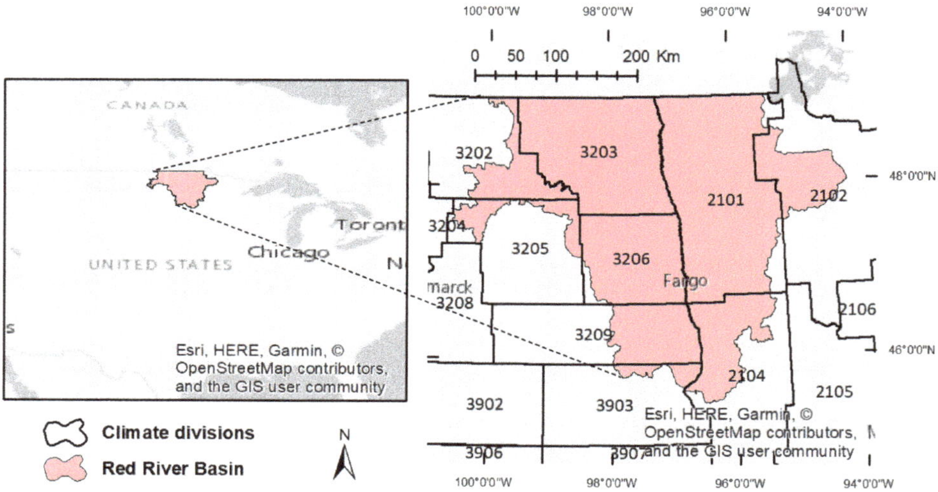


**Figure 1.** Red River Basin (RRB) and its associated climate divisions. CDs 3203, 3206, and 2101 are entirely within the boundaries of the RRB.

To overcome the high covariance of these variables, they are standardized by:

$$P_{i,j}^{stz} = \frac{P_{i,j} - \mu_j^P}{\sigma_j^P} \quad (1)$$

$$SP_{i,j}^{stz} = \frac{SP_{i,j} - \mu_j^{SP}}{\sigma_j^{SP}} \quad (2)$$

$$R_{i,j}^{stz} = \frac{R_{i,j} - \mu_j^R}{\sigma_j^R} \quad (3)$$

$$SWS_{i,j}^{stz} = \frac{SWS_{i,j} - \mu_j^{SWS}}{\sigma_j^{SWS}} \quad (4)$$

where $P_{i,j}^{stz}$, $SP_{i,j}^{stz}$, $R_{i,j}^{stz}$, and $SWS_{i,j}^{stz}$ are the standardized values of precipitation, snowpack, surface runoff, and soil water storage of grid $j$ in month $i$, respectively; $P_{i,j}$, $SP_{i,j}$, $R_{i,j}$, and

$SWS_{i,\,j}$ are the precipitation, snowpack, surface runoff, and soil water storage of grid $j$ in month $i$, respectively; $\mu_j^P$, $\mu_j^{SP}$, $\mu_j^R$, and $\mu_j^{SWS}$ are the mean values of precipitation, snowpack, surface runoff, and soil water storage of grid $j$ in the entire period, respectively; and $\sigma_j^P$, $\sigma_j^{SP}$, $\sigma_j^R$, and $\sigma_j^{SWS}$ are the standard deviation values of precipitation, snowpack, surface runoff, and soil water storage of grid $j$ in the entire period, respectively. By performing a correlation-based R-mode principal component analysis (PCA) [36] on the standardized variables, the SHADI is estimated for grid $j$ and month $i$. The principal components (PCs) are provided by [36]:

$$\begin{bmatrix} PC_{i,\,j}^1 \\ PC_{i,\,j}^2 \\ PC_{i,\,j}^3 \\ PC_{i,\,j}^4 \end{bmatrix} = \begin{bmatrix} a_i^{11} & a_i^{12} & a_i^{13} & a_i^{14} \\ a_i^{21} & a_i^{22} & a_i^{23} & a_i^{24} \\ a_i^{31} & a_i^{32} & a_i^{33} & a_i^{34} \\ a_i^{41} & a_i^{42} & a_i^{43} & a_i^{44} \end{bmatrix} \cdot \begin{bmatrix} P_{i,\,j} \\ SP_{i,\,j} \\ R_{i,\,j} \\ SWS_{i,\,j} \end{bmatrix} \tag{5}$$

where ($PC_{i,\,j}^1$) to ($PC_{i,\,j}^4$) are the first to fourth PCs for grid $j$ in month $i$; and $a_i^{11}$ to $a_i^{14}$, $a_i^{21}$ to $a_i^{24}$, $a_i^{31}$ to $a_i^{34}$, and $a_i^{41}$ to $a_i^{44}$ are, respectively, the normalized eigenvectors of the first, second, third, and fourth PC in month $i$. The elements of the eigenvector are unequal weights in the PCs for month $i$.

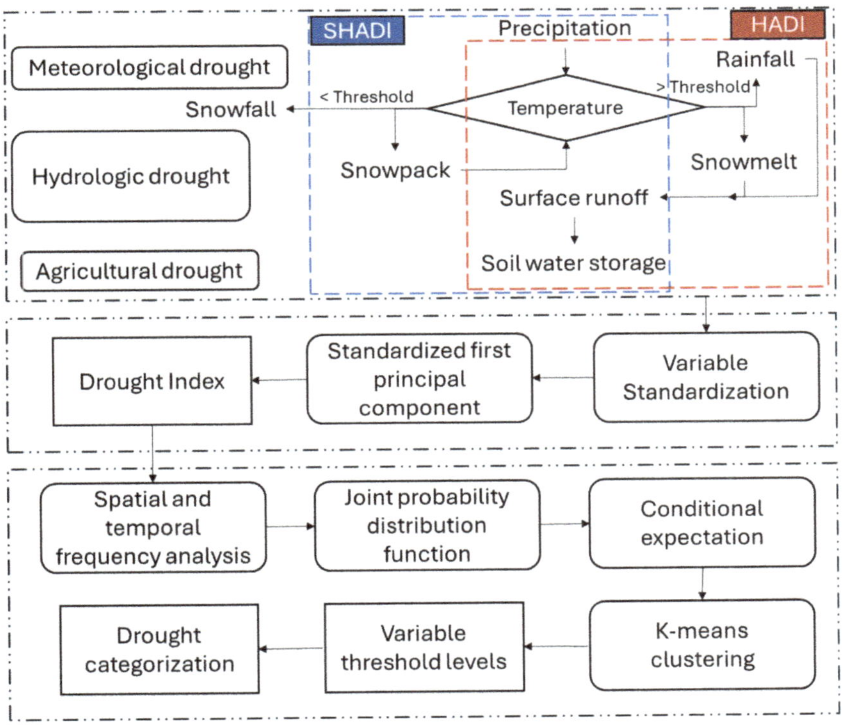


**Figure 2.** Flowchart for calculating HADI and SHADI and for categorizing drought.

The PCs should be standardized to be independent of time and space and, thus, comparable for any normal conditions. The mean and standard deviation values of the first PCs throughout the study period in each grid are calculated. The SHADI is computed by standardization of the first PCs for each grid and month or the ratio of the difference between each first PC value and its mean to its standard deviation:

$$SHADI_{i,\,j} = \frac{PC^{1}_{i,\,j} - \mu_{PC^{1}_{j}}}{\sigma_{PC^{1}_{j}}} \tag{6}$$

where $SHADI_{i,j}$ is the SHADI of grid $j$ in month $i$; $PC^{1}_{i,\,j}$ is the first PC of grid $j$ in month $i$; and $\mu_{PC^{1}_{j}}$ and $\sigma_{PC^{1}_{j}}$ are the mean and standard deviation of the first PC of grid j. The first PC is selected for estimation of the SHADI, since a high proportion of information is found in the first PC.

To categorize droughts, a customized drought categorization based on variable threshold levels [27] is implemented. According to the range of the SHADI values, they are divided into a defined number of classes. To estimate the spatial and temporal frequencies of the SHADI, a cell-by-cell-based analysis is performed. The frequency of each drought class during the study period is estimated for each cell, and the temporal frequencies of all drought classes for all cells are also estimated. The frequency and class of droughts form a bivariate function. Thus, a joint probability distribution function (JPDF) is determined. To find the probability of the occurrence of each class of drought, the conditional expectation is calculated. The lower the conditional expectation, the lower the probability of occurrence and the higher risk of rare drought. The conditional expectation for each class of drought is provided by [37]:

$$g(f|c) = \frac{g(f,c)}{g(c)} = \begin{bmatrix} g(0|1) & g(0|2) & \ldots & g(0|n) \\ g(1|1) & g(1|2) & \ldots & g(1|n) \\ \ldots & \ldots & \ldots & \ldots \\ g(m|1) & g(m|2) & \ldots & g(m|n) \end{bmatrix} \tag{7}$$

in which:

$$g(f,c) = \begin{bmatrix} g(0,1) & g(0,2) & \ldots & g(0,n) \\ g(1,1) & g(1,2) & \ldots & g(1,n) \\ \ldots & \ldots & \ldots & \ldots \\ g(m,1) & g(m,2) & \ldots & g(m,n) \end{bmatrix} \tag{8}$$

$$g(c) = \sum\nolimits_{c=1}^{n} \sum\nolimits_{f=1}^{m} g(f,c) \tag{9}$$

where $g(f,c)$ is the JPDF of frequency $f$ and drought class $c$; $g(c)$ is the probability distribution function of drought class $c$; and $g(f \mid c)$ is the conditional probability. The conditional expectation of different frequencies for each associated class, $P_c$ or $E(f \mid c)$ can be expressed as:

$$P_c = E(f|c) = \sum\nolimits_{f=1}^{m} (f \cdot g(f|c)) \tag{10}$$

To derive the threshold levels for different drought categories, the *k*-means clustering method [38] is used. The *k*-means clustering employs a partitioning approach in which the observations are separated into n clusters without using a hierarchical method. Partitioning is the most commonly used method among nonhierarchical methods of clustering. The specific procedures for determining the variable threshold levels using *k*-means include [38]: (1) select n items to serve as seeds (i.e., initial cluster centroids); (2) assign each observation in the dataset to the cluster with the nearest seed/centroid based on the squared Euclidean distance; (3) recalculate the centroid for each cluster (i.e., the mean vector of all observations in the cluster); and (4) repeat steps 2 and 3 until no observations move to different clusters.

To account for the slow emergence and recession of drought, Svoboda et al. (2002) used six categories (exceptional, extreme, severe, moderate, abnormal, and normal) in the U.S. drought monitor (USDM) [39]. The same numbers of drought categories in the USDM are used in this study. Four wet classes (including exceptional, extreme, severe, and moderate wet) are also considered to categorize the wet conditions. Thus, the customized drought categorization is based on these ten dry and wet clusters. Ten points that are mutually farthest apart are selected as initial seeds in the *k*-means clustering. The observation vector can be expressed as:

$$\begin{cases} y_1 = (c_1, P_1) \\ y_2 = (c_2, P_2) \\ \vdots \\ y_n = (c_n, P_n) \end{cases} \tag{11}$$

where $c_1$ is representative of the lower class limit for class 1 of the SHADI; and $P_1$ is the expected value of probability of occurrence of class 1. The *k*-means method clusters these n observations into ten groups. The initial centroids are the ten observations that are farthest apart. The procedure starts by calculating the squared Euclidean distance between each pair of observations:

$$\begin{cases} d^2(y_k, y_{k+1}) = (c_k - c_{k+1})^2 + (P_k - P_{k+1})^2 \\ d^2(y_k, y_{k+2}) = (c_k - c_{k+2})^2 + (P_k - P_{k+2})^2 \\ \vdots \\ d^2(y_{k+1}, y_{k+2}) = (c_{k+1} - c_{k+2})^2 + (P_{k+1} - P_{k+2})^2 \\ \vdots \\ d^2(y_{n-2}, y_n) = (c_{n-2} - c_n)^2 + (P_{n-2} - P_n)^2 \\ d^2(y_{n-1}, y_n) = (c_{n-1} - c_n)^2 + (P_{n-1} - P_n)^2 \end{cases} \tag{12}$$

where $d^2(y_k, y_{k+1})$ is the squared Euclidean distance between $y_k$ and $y_{k+1}$ observation vectors for k (k = 1, . . . , n). Ten points with the longest pairwise distances are selected as the initial seeds. The distances from each observation to the initial seeds are calculated. Then, the centroid of each cluster and the distance from each observation to each centroid are calculated. These processes are repeated until the cluster assignments in the last step are the same as those in the previous step.

### 2.3. Assessment of SHADI in Identification and Categorization of Droughts

To assess the performance of the SHADI in identification of droughts, it was compared with the HADI and USDM. After the assessment of the SHADI, the new customized drought categorization was applied to the RRB. The results based on the SHADI and HADI were further compared, and their performances were evaluated for the entire RRB and separately for CDs 2101, 3203, and 3206 (Figure 1).

The SHADI, HADI, and USDM (Svoboda et al., 2002; USDM 2019) were compared in the form of drought severity and coverage index (DSCI) [40]. The DSCI turned the USDM categorical drought into one value. The DSCI is used to compare both severity and coverage of droughts in the USDM, SHADI, and HADI. The *DSCI* is provided by [40]:

$$DSCI = \sum_{i=0}^{4} (i+1) \times ACD(i) \tag{13}$$

where *ACD* (*i*) is the percentage of area coverage of drought in category *i*. The USDM (Svoboda et al., 2002) is "a composite product" based on the Objective Drought Indicator (ODI) Blends. The USDM product is a blend of six main drought indicators and other objectively selected ancillary indices, including the Palmer drought severity index (PDSI) [41], CPC soil moisture model (percentiles), USGS weekly streamflow (percentiles), percent of normal, SPI, and satellite vegetation [39]. Additional information about the conditions of soil moisture and snowpack is added by other ancillary indices (e.g., SWSI).

There are some differences in the derivations of USDM, HADI, and SHADI. Soil moisture and observed streamflow are used in the USDM, while the simulated surface runoff and soil water storage in the root zone are used in the SHADI and HADI. Although precipitation and snowpack are considered in the derivation of the USDM products, the separation of rainfall and snowfall, on the one hand, and snowmelt are ignored in the USDM. In contrast, the SHADI and HADI account for these dominant hydroclimatic processes in cold climate regions (precipitation and snowpack in the SHADI and rainfall and snowmelt in the HADI).

### 2.4. Cluster Analysis

*k*-means clustering was applied to define dry and wet conditions. The outputs of the cluster analysis for precipitation, rainfall, snowpack, snowmelt, surface runoff, and soil water storage (i.e., the inputs for HADI and SHADI) were compared with their corresponding indices. Cluster analysis is a multivariate technique that identifies an optimal grouping, where each cluster contains observations that are as similar as possible. A nonhierarchical *k*-means clustering method [38] was used in this study.

In *k*-means clustering, the number of clusters must be specified. For the hydroclimatic variables in this study, only two clusters (dry and wet) were defined to avoid unnecessary complexity.

### 2.5. Correlation Analysis

Pearson and Spearman correlations are widely used statistical methods to measure the strength and direction of the relationship between two variables. The Pearson correlation coefficient assesses the linear relationship between variables, assuming that the data are normally distributed and have a linear association. It provides values between −1 and 1, where values close to 1 or −1 indicate a strong positive or negative linear relationship, respectively, and values near 0 indicate no linear relationship. In contrast, the Spearman rank correlation coefficient is a non-parametric measure that evaluates the monotonic relationship between variables, making it suitable for data that do not meet the assumptions of normality or linearity. Spearman correlation ranks the data and calculates the correlation based on the ranked values, making it robust against outliers and useful for analyzing non-linear associations. Both correlations were used in this study to evaluate the agreement between the indices (the HADI and SHADI).

## 3. Results and Discussion

### 3.1. Categorization and Classification of Droughts by SHADI and HADI

The droughts in the study period were characterized by using the customized drought categorization method for both the SHADI and HADI. The threshold levels for different categories of drought based on the SHADI and HADI were close (Table 1). The differences in variable threshold levels between the HADI and SHADI are minimal, with a negligible value of 0.1. The lower limits for the categories based on the SHADI are as follows: exceptional wet (4.7), normal condition (−1.2), abnormal drought (−1.5), moderate drought (−1.6), severe drought (−1.8), and extreme drought (−1.9). In comparison, the lower limits for the same categories based on the HADI are exceptional wet (4.6), normal condition (−1.1), abnormal drought (−1.4), moderate drought (−1.5), severe drought (−1.7), and extreme drought (−1.8). Similarly, the upper limits for the SHADI are abnormal drought (−1.2), moderate drought (−1.5), severe drought (−1.6), extreme drought (−1.8), and exceptional drought (−1.9). For the HADI, the corresponding upper limits are abnormal drought (−1.1), moderate drought (−1.4), severe drought (−1.5), extreme drought (−1.7), and exceptional drought (−1.8). Since the study period and the study area remained unchanged, the threshold levels did not alter. This similarity can confirm the variability of the derived threshold levels by time and geographic locations.

**Table 1.** Variable threshold levels and drought categories based on the SHADI and HADI.

| SHADI Values | HADI Values | Category | Symbol |
|---|---|---|---|
| **SHADI > 4.7** | HADI > 4.6 | Exceptional wet | W4 |
| **2.9 < SHADI < 4.7** | 2.9 < HADI < 4.6 | Extreme wet | W3 |
| **1.3 < SHADI < 2.9** | 1.3 < HADI < 2.9 | Severe wet | W2 |
| **0.2 < SHADI < 1.3** | 0.2 < HADI < 1.3 | Moderate wet | W1 |
| **−1.2 < SHADI < 0.2** | −1.1 < HADI < 0.2 | Normal condition | N |
| **−1.5 < SHADI < −1.2** | −1.4 < HADI < −1.1 | Abnormal drought | D0 |
| **−1.6 < SHADI < −1.5** | −1.5 < HADI < −1.4 | Moderate drought | D1 |
| **−1.8 < SHADI < −1.6** | −1.7 < HADI < −1.5 | Severe drought | D2 |
| **−1.9 < SHADI < −1.8** | −1.8 < HADI < −1.7 | Extreme drought | D3 |
| **SHADI < −1.9** | HADI < −1.8 | Exceptional drought | D4 |

### 3.2. Comparison of the Results of SHADI and HADI in the RRB

Figure 3 shows the comparison of temporal and spatial distributions in the HADI and SHADI. Notable dissimilarities were observed in the cold seasons and in the dry–wet or wet–dry transitions (Figure 3a). The maximum difference between the HADI (1.73) and SHADI (−0.46) was observed in April 2004, featuring a transition from wet to dry conditions. Another transition period from dry to wet occurred in January 2006 (HADI = −0.46; SHADI = 0.23). In this month, the HADI remained negative, while the SHADI preidentified the upcoming wet condition. Moreover, in the middle of the long dry period (2004–2006), the HADI (1.41) identified a high-intensity wet condition in February 2005 (Figure 3c,e). In contrast, the SHADI (0.47) showed a wet condition with considerably less intensity in this month (Figure 3b,d), indicating that the SHADI exhibited fewer fluctuations than the HADI.

Table 2 presents the statistical properties of the SHADI and HADI. A slightly higher maximum value was observed for the SHADI (2.01), while the HADI exhibited a lower minimum value (−0.87). However, the average and standard deviation of the two indices were quite similar, indicating comparable overall variability and central tendencies. The Pearson and Spearman correlation coefficients of the SHADI and HADI were 0.75 and 0.78, respectively. The Spearman correlation coefficient indicated that the intensities of dry or wet conditions identified by the HADI and SHADI and their corresponding ranks among different months in the study period were close in the warm seasons. The Pearson correlation coefficient also showed the similarity between the HADI and SHADI values in the warm seasons. In fact, the differences were negligible from May to September in all years.

**Table 2.** Statistical properties of SHADI and HADI.

| | SHADI | | HADI |
|---|---|---|---|
| Max | 2.01 | | 1.94 |
| Min | −0.85 | | −0.87 |
| Average | 0 | | 0 |
| Standard deviation | 0.70 | | 0.69 |
| Pearson correlation | | 0.75 | |
| Spearman correlation | | 0.78 | |

The analysis of the dissimilarities in the spatial distributions of the HADI and SHADI helps determine the sensitivity of the indices to drought drivers (e.g., temperature-driven or precipitation-driven droughts). There was a minor difference between the HADI and SHADI for 2003 (Figure 4a,f) and 2007 (Figure 4e,j). However, major differences were observed between the HADI and SHADI in 2004 (Figure 4b,g), 2005 (Figure 4c,h), and 2006 (Figure 4d,i). These significant differences were in the upper RRB in 2004, the lower RRB in 2005, as well as the upper and northeast part of the RRB in 2006.

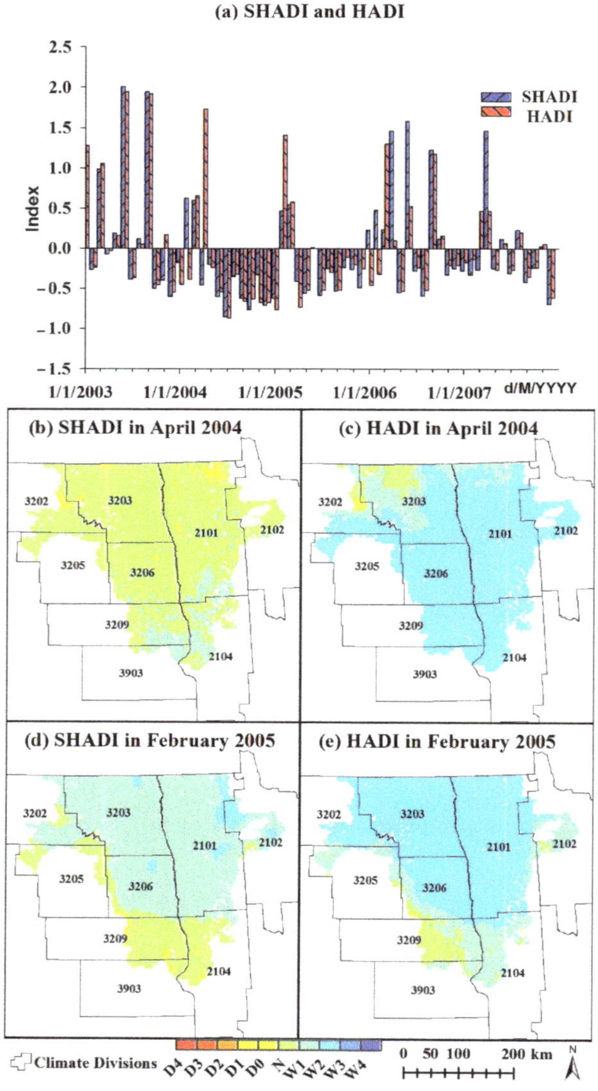


**Figure 3.** Comparison of temporal and spatial distributions of HADI and SHADI.

As shown in Figure 5, the HADI and SHADI had similar values in 2003 (Figure 5a,f) and 2007 (Figure 5e,j). However, the SHADI indicated more intense droughts in the upper RRB in 2003 (Figure 5b,g) and in the lower RRB in 2004 (Figure 5c,h). By the end of the 2004–2006 drought (Figure 5d,i), the SHADI identified a wetter condition in the upper and northeast part of the RRB. The lowest annual precipitation occurred in 2003 (Figure 5k) and 2006 (Figure 5n). In addition, the western side of the RRB had a lower amount of precipitation than that of the eastern side. Figure 5p–t show the spatial distribution of the mean annual temperatures from 2003 to 2007. The lowest and the highest temperatures occurred in 2004 and 2006, respectively. The year 2004 was the coldest in the study period. However, the high temperature in the upper RRB caused rainfall instead of snow. Thus, the SHADI identified a dry condition with higher drought intensity due to the lack of snowpack (Figure 5b,g). Both precipitation and temperature in the lower RRB in 2005 were relatively low, and the intensity of drought based on the SHADI was higher (Figure 5c,h). The low precipitation and high temperature in the upper and northeast part of the RRB in 2006 led to high-intensity drought based on the HADI (Figure 5d,i). Therefore, it can be concluded that the SHADI is more sensitive to low temperatures, and the intensity of droughts identified by the SHADI can be higher in cold climate conditions.

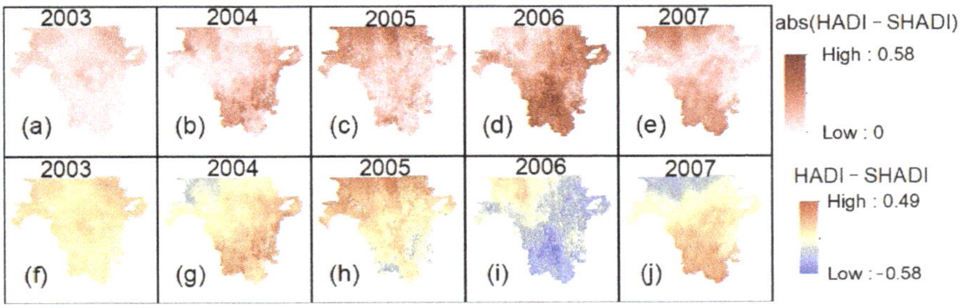


**Figure 4.** Differences between HADI and SHADI in the RRB. (**a–e**) Absolute differences between HADI and SHADI values from 2003 to 2007. (**f–j**) Raw differences between HADI and SHADI values from 2003 to 2007.

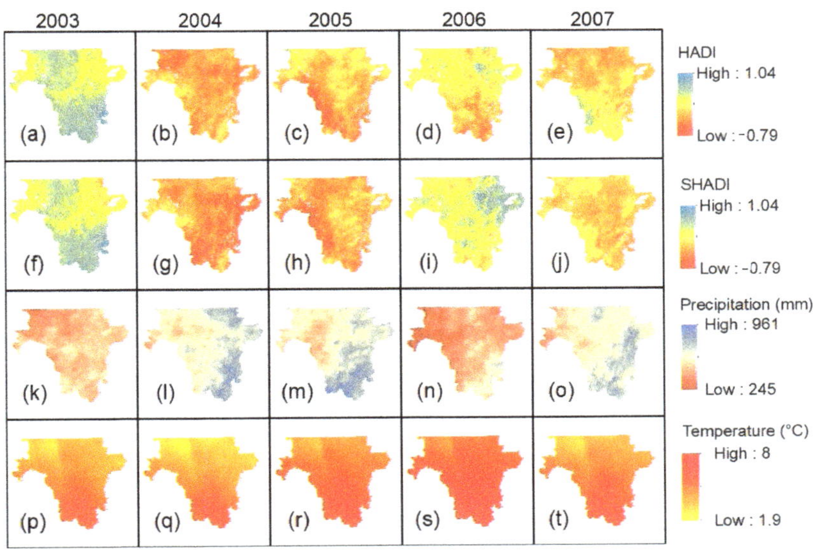


**Figure 5.** Spatial and temporal distributions of HADI, SHADI, temperature, and precipitation in the RRB. (**a–e**) Spatial distribution of HADI from 2003 to 2007. (**f–j**) Spatial distribution of SHADI from 2003 to 2007. (**k–o**) Spatial distribution of precipitation from 2003 to 2007. (**p–t**) Spatial distribution of temperature from 2003 to 2007.

### *3.3. Comparison of SHADI, HADI, and USDM in the Form of DSCI*

To compare the DSCIs of the SHADI, HADI, and USDM, the dry span in the study period was divided into three short dry spells: (1) from the end of 2003 to the end of 2004,

(2) from the beginning of 2005 until the end of 2005, and (3) from the beginning of 2006 to the beginning of 2007. As shown in Figure 6, despite the differences in the values of the DSCI of the USDM, SHADI, and HADI, they identified the first dry spell at the same time in all CDs. The DSCIs of the USDM were higher than those of the HADI in all CDs. In contrast, the DSCIs of the SHADI were higher than those of the USDM in CDs 2101 and 3203. The peak DSCIs of the HADI and SHADI occurred with a shorter delay than the DSCI of the USDM in the second dry spell. The DSCI of the SHADI was higher than those of the USDM and HADI in CDs 2101 and 3203. The DSCIs of the USDM and HADI were close to or higher than the DSCI of the SHADI in CD 3206. The DSCI of the USDM was noticeably higher than those of the SHADI and HADI in the third dry spell.

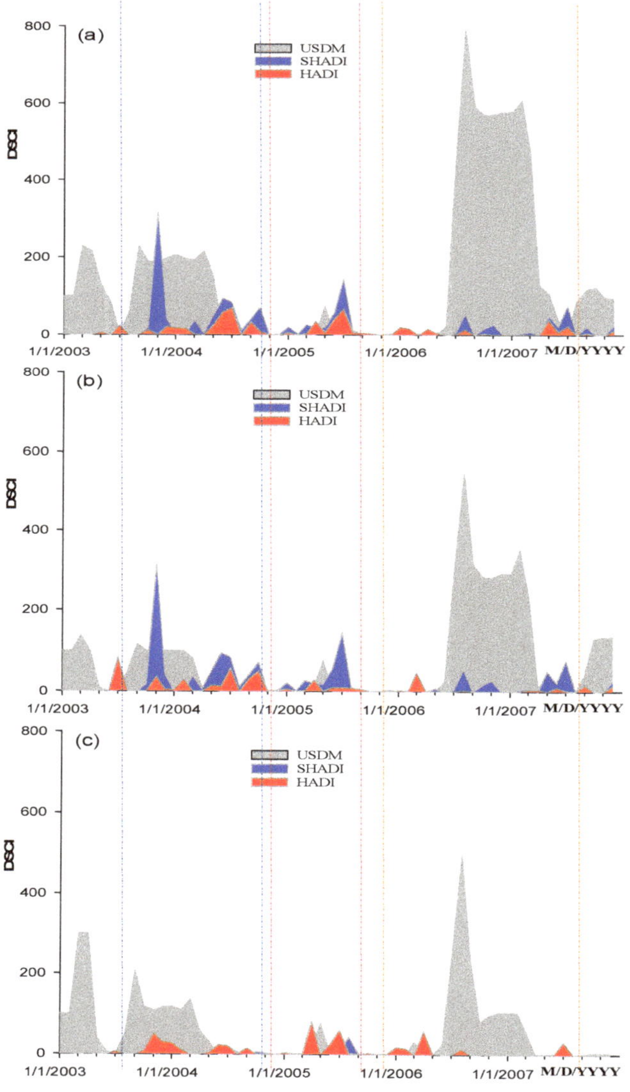


**Figure 6.** Severity and coverage drought index of the USDM, SHADI, and HADI in climate divisions (**a**) 2101, (**b**) 3203, and (**c**) 3206.

The dissimilarities among the results of the SHADI, HADI, and USDM can be associated with three main reasons. First, different numbers and types of inputs (hydroclimatic variables as drought indicators or drought indices) were used for the derivation of these indices. The SHADI accounted for snowpack and snow drought, and the HADI considered snowmelt and rainfall. However, the USDM did not consider the separation of rainfall and snowfall, snow accumulation, and snowmelt as the dominant hydroclimatic processes in cold climate regions. Second, the approach of one-weight-fits-all can be a limitation of the USDM in weighting coefficients of the involved indices and in the representation of seasonal changes. However, by using the PCA, the weighting coefficients for different hydroclimatic variables in the HADI were derived based on the employed time series instead of the fixed coefficients used in the USDM. Third, in this research, the variable threshold levels were used in drought categorization for the HADI and SHADI instead of the fixed threshold levels based on the percentages for the USDM. Since a dry period was intentionally chosen to test the SHADI and HADI, these discrepancies in the results were expected. The main purpose for the derivation and utilization of variable threshold levels in drought categorization is to fulfil the necessity of consideration of both spatial and temporal distributions of droughts [42]. Hence, different thresholds for each drought category would be expected if different periods were selected. Therefore, the properties of the HADI and SHADI, which were incorporated to improve drought identification and categorization, resulted in these discrepancies among the indices.

### 3.4. *Cluster Analyses of the Inputs of HADI and SHADI*

Figure 7 shows the HADI and SHADI and the clustering analysis results based on their associated inputs in CD 2101 (Figure 7a,b), CD 3203 (Figure 7c,d), and CD 3206 (Figure 7e,f). The SHADI and cluster analysis identified the same dry/wet conditions in 47%, 72%, and 68% of times for CDs 2101, 3203, and 3206, respectively. Results of the HADI and cluster analysis for identification of dry/wet conditions agreed in 43%, 53%, and 52% of times for CDs 2101, 3203, and 3206, respectively.

In CD 2101, the first dry period occurred from fall 2003 to the end of winter 2004. In this dry period, the monthly temperature was lower than the mean monthly temperature in CD 2101, especially in the cold seasons. As a result, the precipitation was in the form of snowfall, and the snowpack increased over time, reaching its capacity by the end of the cold season in February 2004. Then, the snowmelt process was initiated by a temperature increase in March 2004. The snowmelt-induced surface runoff and soil water storage increased with a one-month delay. The SHADI identified this dry period two months earlier than the HADI (in October 2003). In the second dry period, the monthly temperature was higher than the mean monthly temperature in the cold seasons. Therefore, the majority of precipitation was in the form of rainfall. Although the rainfall was higher than that in the other years in the study period, both HADI and SHADI identified a dry condition in this period. The snowpack and snowmelt were relatively low. The drought identified by the HADI can be justified by the low snowmelt in this period, since there is certain relationship between the HADI and snowmelt. However, there was no direct correlation between the SHADI and temperature, precipitation, snowpack, surface runoff, and soil water storage. This can be verified by the similar loadings of the PCs or the weighting coefficients of all inputs. Thus, the SHADI represented a combination of all hydroclimatic variables. Due to the relatively low values of these variables, the SHADI identified a dry condition in this period. The third period was a short dry period from the end of 2006 to March 2007. In this period, both the HADI and SHADI identified a dry spell with below-normal precipitation and soil water storage. The deficit in the antecedent precipitation during 2006 led to a decrease in the available soil moisture. Thus, this drought can be categorized as a classical rainfall deficit drought.

The drought intensity based on the SHADI was higher than that of the HADI in CD 3203 in November 2003. This can be justified by the lower temperature in CD 3203 than that in CD 2101 and CD 3206. In addition, the SHADI exhibited more consistency in the second dry period than the HADI in CD 3203. For example, the HADI identified a wet condition in October 2005 (middle of the second dry period) due to an increase in snowmelt, while the SHADI remained negative in this month.

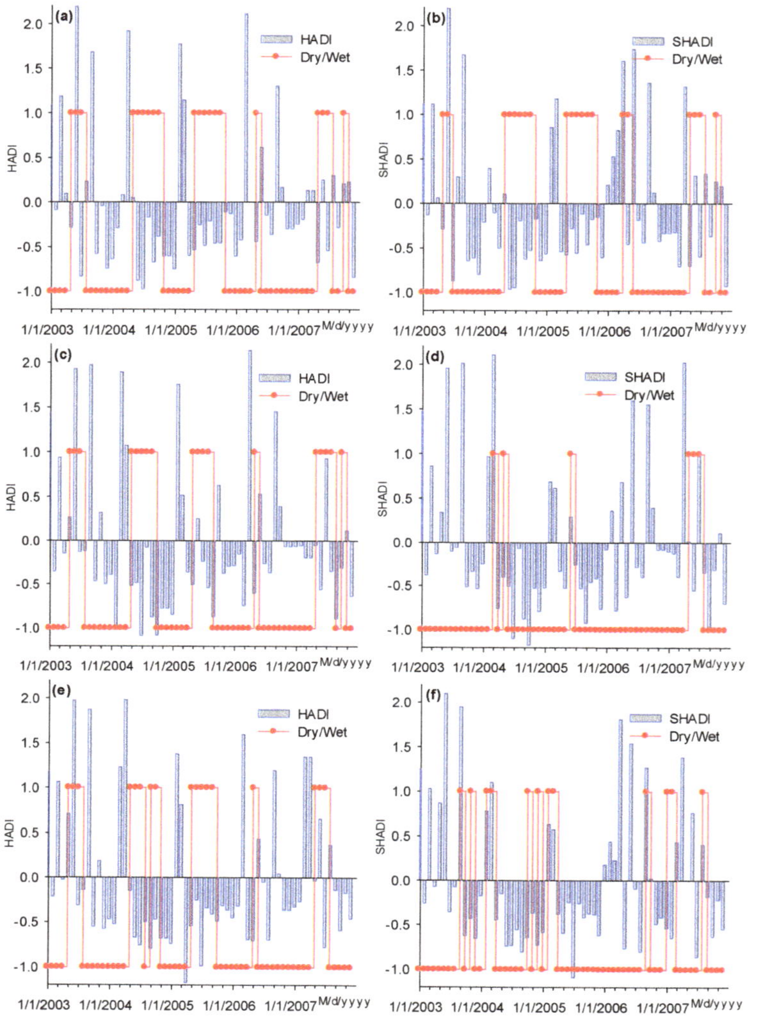


**Figure 7.** Clustering results based on the inputs of the HADI and SHADI and their comparisons with their associated indices in climate divisions (**a**,**b**) 2101, (**c**,**d**) 3203, and (**e**,**f**) 3206.

Two types of major differences were observed between the HADI and SHADI in CD 3206. First, in some months, the HADI values were positive, while the SHADI values were negative. For instance, the HADI values were 0.19 and 0.77 in November 2003 and February 2004, respectively. However, the SHADI values were −0.43 and −0.52. Compared with the HADI (1.98), which showed a high-intensity wet condition in April 2004, the SHADI (−0.44) identified a dry condition in this month, which can be interpreted as a short-term lead prediction of the forthcoming dry period by the SHADI. Second, in some months close to the end of the 2004–2006 dry period, the SHADI turned to positive values, whereas the HADI remained negative. For example, the SHADI values were 0.18, 0.44, and 1.81 in

January, February, and April 2006, respectively. However, the HADI values were −0.45, −0.31, and −0.69 for these months. In fact, by the end of this long dry period, the SHADI preidentified a wet condition in January 2006, two months prior to the HADI, since SHADI is based on snowpack and precipitation rather than available water, such as rainfall and snowmelt, used in the HADI.

The longest similarity between the indices and the clustering results on their inputs was observed from January to December of 2005 in CD 3206 for the SHADI (Figure 7f). Thus, the results of the SHADI were closer to those from the cluster analysis especially for CD 3203. It can be concluded that despite some discrepancies, the results of cluster analysis verified the results of the HADI and SHADI in most cases.

## 4. Conclusions

To enhance drought identification in cold climate regions and assess the impacts of precipitation, snowpack, and snowmelt in drought analysis, a new snow-based hydroclimatic aggregate drought index (SHADI) was developed. It was applied to the RRB, and the results were compared with those of the HADI. Discrepancies were specifically observed between these two drought indices in cold seasons and in dry–wet or wet–dry transitions, which can be attributed to the effects of consideration of available water (rainfall and snowmelt) in the HADI and stored water sources (precipitation and snowpack) in the SHADI. Cluster analysis was performed as a benchmark to compare the results of the HADI and SHADI. The SHADI outperformed the HADI in aligning with the cluster analysis results, matching the identified dry/wet conditions 47%, 72%, and 68% of the time for three selected CDs (2101, 3203, and 3206, respectively). Thus, the SHADI is particularly suitable for identification of snow droughts.

The anomalous snowpack and snow-based drought indices can provide early warning of droughts. Accounting for the snowpack impact in the SHADI facilitated short-term lead predictions of droughts for two months, which was also demonstrated by Abel's et al. (2018) [28,42]. Thus, the SHADI can be used for identifying snow droughts and providing short-term lead predictions of upcoming droughts to mitigate the drought impacts.

As a result of the use of the customized drought categorization, the SHADI and HADI yielded similar DSCI values, which were different from those of the USDM. These discrepancies can be attributed to the incorporation of effective hydroclimatic variables in cold climate regions, implementation of different methods for derivation of weighting coefficients, and application of the customized drought categorization approach and variable threshold levels in the SHADI and HADI for drought identification and categorization.

Precipitation, snowpack, surface runoff, and soil water storage were used to monitor droughts in this study. Other alternative water sources, such as snow cover or SWE, could be explored in future studies. Only the first PC was used for derivation of the SHADI, as it accounted for the highest proportion of variance. However, in the second and third quarters of the year (i.e., spring and summer), the first eigenvalue percentages were lower due to collinearity. Collinearity occurs when some independent variables are highly correlated, which can affect the representation of drought conditions. Thus, it is important to note that the first PC alone may not fully capture the complexities of drought conditions. As discussed by Bazrkar et al. (2020) [27], relying solely on the first PC can overlook important drought dynamics that may be better represented by other components.

The RRB was selected as the study area due to its snow-dominated, cold climate characteristics. The broader applicability of the SHADI to other cold climate regions can be explored in future research. Additionally, future studies could evaluate the SHADI by comparing it with other existing drought indices beyond the HADI and the USDM. Such analyses would offer a more comprehensive understanding of the SHADI's performance and its potential utility across diverse regions and climatic conditions, further assessing its relevance and effectiveness in capturing drought dynamics in snow-dominated, cold climate areas. The GHM model was chosen for this study, as it was the available tool specifically designed to simulate hydroclimatic processes in depression-dominated, cold climate regions like the RRB. However, it does not account for certain snow processes such as snowpack sublimation and redistribution. To address these limitations, Tahmasebi Nasab and Chu (2020) [43] developed an enhanced version, Macro-HyProS, which incorporated a snowfall correction factor for sublimation and redistribution losses. The SHADI can be utilized for real-time monitoring and short-term lead prediction of droughts, providing valuable information to help mitigate the impacts of droughts.

**Author Contributions:** M.H.B.: conceptualization, methodology, software, validation, formal analysis, investigation, writing—original draft, writing—review and editing, visualization. N.Z.: validation, formal analysis, investigation, writing—review and editing, visualization. X.C.: conceptualization, methodology, investigation, writing—review and editing, data curation, supervision, project administration, funding acquisition. All authors have read and agreed to the published version of the manuscript.

**Funding:** This material is based upon work supported by the National Science Foundation under Grant No. NSF EPSCoR Award IIA-1355466. The North Dakota Water Resources Research Institute also provided partial financial support in the form of a graduate fellowship for the first author.

**Institutional Review Board Statement:** Not applicable.

**Informed Consent Statement:** Not applicable.

**Data Availability Statement:** The datasets for this research are available at the NOAA's National Centers for Environmental information (NOAA's NCEI) "https://www.ncdc.noaa.gov/cag/ (accessed on 9 November 2019)" and the United States Drought Monitor (USDM) (https://droughtmonitor.unl.edu/Data/Timeseries.aspx1 (accessed on 27 June 2019).

**Conflicts of Interest:** The authors declare no conflicts of interest.

## References

1. Marty, C.; Schlögl, S.; Bavay, M.; Lehning, M. How much can we save? Impact of different emission scenarios on future snow cover in the Alps. *Cryosphere* **2017**, *11*, 517–529. [CrossRef]
2. Sproles, E.A.; Roth, T.R.; Nolin, A.W. Future snow? A spatial-probabilistic assessment of the extraordinarily low snowpacks of 2014 and 2015 in the Oregon Cascades. *Cryosphere* **2017**, *11*, 331–341. [CrossRef]
3. Huning, S.L.; Aghakouchak, A. Mountain snowpack response to different levels of warming. *Proc. Natl. Acad. Sci. USA* **2018**, *115*, 10932–10937. [CrossRef]
4. National Integrated Drought Information System. Snow Drought: Current Conditions and Impacts in the West. Drought.gov. 13 January 2022. Available online: https://www.drought.gov/drought-status-updates/snow-drought-current-conditions-and-impacts-west-5 (accessed on 27 June 2019).
5. Climate Signals. Western U.S. snow drought: Winter 2018. Climate Signals. 2018. Available online: https://www.climatesignals.org/events/western-us-snow-drought-winter-2018 (accessed on 27 June 2019).
6. Schneider, S.R.; Matson, M. Satellite observations of snowcover in the Sierra Nevadas during the great California drought. *Remote Sens. Environ.* **1977**, *6*, 327–334. [CrossRef]
7. Wiesnet, D. *Winter Snow Drought*; Eos, Transactions American Geophysical Union: Washington, DC, USA, 1981; Volume 62, p. 137.
8. American Meteorological Society. Snow Drought. Glossary of Meteorology. 2019. Available online: http://glossary.ametsoc.org/wiki/Snow_drought (accessed on 27 June 2019).
9. Harpold, A.; Brooks, P.; Rajagopal, S.; Heidbuchel, I.; Jardine, A.; Stielstra, C. Changes in snowpack accumulation and ablation in the intermountain west. *Water Resour. Res.* **2012**, *48*, W11501. [CrossRef]
10. Van Loon, A.F.; Ploum, S.W.; Parajka, J.; Fleig, A.K.; Garnier, E.; Laaha, G.; Van Lanen, H. Hydrological drought types in cold climates: Quantitative analysis of causing factors and qualitative survey of impacts. *Hydrol. Earth Syst. Sci.* **2015**, *19*, 1993–2016. [CrossRef]
11. Cooper, M.G.; Nolin, A.W.; Safeeq, M. Testing the recent snow drought as an analog for climate warming sensitivity of Cascades snowpacks. *Environ. Res. Lett.* **2016**, *11*, 084009. [CrossRef]
12. Harpold, A.A.; Dettinger, M.; Rajagopal, S. *Defining Snow Drought and Why It Matters*; EOS, 98; Eos, Transactions American Geophysical Union: Washington, DC, USA, 2017. [CrossRef]
13. Dierauer, J.R.; Allen, D.M.; Whitfield, P.H. Snow drought risk and susceptibility in the western United States and southwestern Canada. *Water Resour. Res.* **2019**, *55*, 3076–3091. [CrossRef]

14. Mote, P.W. Trends in snow water equivalent in the Pacific Northwest and their climatic causes. *Geophys. Res. Lett.* **2003**, *30*, 1601. [CrossRef]
15. Staudinger, M.; Stahl, K.; Seibert, J.; Clark, M.P.; Tallaksen, L.M. Comparison of hydrological model structures based on recession and low flow simulations. *Hydrol. Earth Syst. Sci.* **2011**, *15*, 3447–3459. [CrossRef]
16. Van Loon, A.F.; Laaha, G. Hydrological drought severity explained by climate and catchment characteristics. *J. Hydrol.* **2015**, *526*, 3–14. [CrossRef]
17. Fang, X.; Pomeroy, W. Snowmelt runoff sensitivity analysis to drought on the Canadian prairies. *Hydrol. Process.* **2007**, *21*, 2594–2609. [CrossRef]
18. Van Loon, A.F.; Van Lanen, H.A.J. A process-based typology of hydrological drought. *Hydrol. Earth Syst. Sci.* **2012**, *16*, 1915–1946. [CrossRef]
19. Van Loon, A.; Laaha, G.; Van Lanen, H.; Parajka, J.; Fleig, A.K.; Ploum, S. Drought and Snow: Analysis of Drivers, Processes and Impacts of Streamflow Droughts in Snow-Dominated Regions. In Proceedings of the American Geophysical Union, Fall Meeting, San Francisco, CA, USA, 18 December 2015.
20. Howitt, R.; MacEwan, D.; Medellín-Azuara, J.; Lund, J.; Sumner, D. Economic Analysis of the 2015 Drought for California Agriculture. UC Davis Center for Watershed Sciences. 2015. Available online: https://watershed.ucdavis.edu/files/biblio/Final_Drought%20Report_08182015_Full_Report_WithAppendices.pdf (accessed on 1 March 2018).
21. Hatchett, B.J.; McEvoy, D.J. Exploring the Origins of Snow Drought in the Northern Sierra Nevada, California. *Earth Interact.* **2018**, *22*, 1–13. [CrossRef]
22. Huning, S.L.; Aghakouchak, A. Global snow drought hot spots and characteristics. *Proc. Natl. Acad. Sci. USA* **2020**, *117*, 19753–19759. [CrossRef]
23. Shafer, B.A.; Dezman, L.E. Development of a surface water supply index (SWSI) to assess the severity of drought conditions in snowpack runoff areas. In Proceedings of the 50th Annual Western Snow Conference, Reno, NV, USA, 19–23 April 1982; pp. 164–175.
24. Keyantash, J.A.; Dracup, J.A. An aggregate drought index: Assessing drought severity based on fluctuations in the hydrologic cycle and surface water storage. *Water Resour. Res.* **2004**, *40*, W09304. [CrossRef]
25. Staudinger, M.; Stahl, K.; Seibert, J. A drought index accounting for snow. *Water Resour. Res.* **2014**, *50*, 7861–7872. [CrossRef]
26. Bazrkar, M.H.; Zhang, J.; Chu, X. Hydroclimatic aggregate drought index (HADI): A new approach for identification and categorization of drought in cold climate regions. *Stoch. Environ. Res. Risk Assess.* **2020**, *34*, 1847–1870. [CrossRef]
27. Abel, D.; Pollinger, F.; Paeth, H. Influence of snow water equivalent on droughts and their prediction in the USA. *Hydrol. Earth Syst. Sci. Discuss.* **2018**. preprint. [CrossRef]
28. Livneh, B.; Badger, A.M. Drought less predictable under declining future snowpack. *Nat. Clim. Chang.* **2020**, *10*, 452–458. [CrossRef]
29. NDAWN. The Drought of the 1930s. 2014. Available online: https://www.ndsu.edu/ndscoblog/?p=626 (accessed on 3 July 2014).
30. NOAA National Centers for Environmental Information. Climate at a Glance: Divisional Haywood. December 2024. Available online: https://www.ncei.noaa.gov/access/monitoring/climate-at-a-glance/divisional/haywood (accessed on 13 December 2024).
31. Guttman, N.B.; Quayle, R.G. A Historical Perspective of U.S. Climate Divisions. *Bull. Am. Meteorol. Soc.* **1996**, *77*, 293–304. [CrossRef]
32. US Geological Survey (2020). Available online: https://waterdata.usgs.gov/nd/nwis/uv?site_no=05054000 (accessed on 3 March 2020).
33. USDM. United States Drought Monitor. 2019. Available online: https://droughtmonitor.unl.edu/Data/Timeseries.aspx (accessed on 27 June 2019).
34. Chu, X.; Lin, Z.; Tahmasebi Nasab, M.; Zeng, L.; Grimm, K.; Bazrkar, M.H.; Wang, N.; Liu, X.; Zhang, X.; Zheng, H. Macro-scale grid-based and subbasin-based hydrologic modeling: Joint simulation and cross-calibration. *J. Hydroinform.* **2019**, *21*, 77–91. [CrossRef]
35. Cattell, R.B. *Factor Analysis: An Introduction and Manual for the Psychologist and Social Scientist*; Greenport Press: Westport, CT, USA, 1952; p. 462.
36. Hogg, R.V.; Craig, A.T. *Introduction to Mathematical Statistics*, 4th ed.; Macmillan Publishing Co., Inc.: New York, NY, USA, 1978.
37. MacQueen, J.B. Some Methods for classification and Analysis of Multivariate Observations. In *Proceedings of the 5th Berkeley Symposium on Mathematical Statistics and Probability, Berkeley, CA, USA, 21 June–18 July 1965*; University of California Press: Berkeley, CA, USA, 1967; Volume 1, pp. 281–297.
38. Svoboda, M.; LeComte, D.; Hayes, M.; Heim, R.; Gleason, K.; Angel, J.; Rippey, B.; Tinker, R.; Palecki, M.; Stooksbury, D.; et al. The drought monitor. *Bull. Am. Meteorol. Soc.* **2002**, *83*, 1181–1190. [CrossRef]
39. Akyuz, F.A. Drought Severity and Coverage Index. United States Drought Monitor. 2017. Available online: https://droughtmonitor.unl.edu/AboutUSDM/AbouttheData/DSCI.aspx (accessed on 27 June 2019).
40. Palmer, W.C. *Meteorological Drought*; Res. Pap. No. 45; U.S. Weather Bureau: Washington, DC, USA, 1965.
41. Mishra, A.K.; Singh, V.P. A review of drought concepts. *J. Hydrol.* **2010**, *391*, 202–216. [CrossRef]

42. Modi, P.A.; Small, E.E.; Kasprzyk, J.; Livneh, B. Investigating the Role of Snow Water Equivalent on Streamflow Predictability during Drought. *J. Hydrometeorol.* **2022**, *23*, 1607–1625. [CrossRef]
43. Tahmasebi Nasab, M.; Chu, X. Macro-HyProS: A new macro-scale hydrologic processes simulator for depression-dominated cold climate regions. *J. Hydrol.* **2020**, *580*, 124366. [CrossRef]

atmosphere

Article

# Investigating Hydrological Drought Characteristics in Northeastern Thailand in CMIP5 Climate Change Scenarios

**Sornsawan Chatklang, Piyapong Tongdeenok * and Naruemol Kaewjampa**

Watershed Management and Environmental Program, Department of Conservation, Faculty of Forestry, Kasetsart University, Bangkok 10900, Thailand; sornsawan.chatkl@ku.th (S.C.); ffornmk@ku.ac.th (N.K.)
* Correspondence: fforppt@ku.ac.th

**Abstract:** In this study, we analyzed the predictions of hydrological droughts in the Lam Chiang Kri Watershed (LCKW) by using the Soil and Water Assessment Tool (SWAT) and streamflow data for 2010–2021. The objective was to assess the streamflow drought index (SDI) for 5-, 10-, 25-, and 50-year return periods (RPs) in 2029 and 2039 in two representative concentration pathway (RCP) scenarios: the moderate climate change scenario (RCP 4.5) and the high-emission scenario (RCP 8.5). The SWAT model showed high accuracy ($R^2$ = 0.82, NSE = 0.78). In RCP4.5, streamflow is projected to increase by 34.74% for 2029 and 18.74% for 2039, while in RCP8.5, a 37.06% decrease is expected for 2029 and 55.84% for 2039. A historical analysis indicated that there were frequent short-term droughts according to SDI-3 (3-month-period index), particularly from 2014 to 2015 and 2020 to 2021, and severe droughts according to SDI-6 (6-month-period index) in 2015 and 2020. The RCP8.5 projections indicate worsening drought conditions, with critical periods from April to June. A wavelet analysis showed that there is a significant risk of severe hydrological drought in the LCKW. Drought characteristic analysis indicated that high-intensity events occur with low frequency in the 50-year RP. Conversely, high-frequency droughts with lower intensity are observed in RPs of less than 50 years. The results of this study highlight an increase in severe drought risk in high emission scenarios, emphasizing the need for water management.

**Keywords:** climate change; CMIP5; hydrological drought characteristics; Streamflow Drought Index (SDI)

## 1. Introduction

Climate change exerts a significantly influence on extreme events [1], particularly droughts [2], which have long-lasting impacts on human life, the environment, industry, and the economy [3]. Hydrological droughts, characterized by below-average streamflow, are crucial issues to consider in water resource planning and management due to increasing demand and population growth [4]. As climate change continues to alter weather patterns, predicting droughts becomes essential to ensuring efficient water resource management, irrigation system operation, agricultural production, and national economic stability [5].

However, predicting hydrological droughts is challenging due to the nonstationary nature of hydrological processes influenced by climate change [5]. These challenges complicate integrated water resource management, as droughts significantly reduce available water resources; therefore, it is necessary to design strategies to balance supply and demand [6]. Moreover, climate change exacerbates drought severity on a global scale, making it imperative to evaluate hydrological drought in various climate scenarios. Human activities, such as water over-extraction and land-use changes, further influence drought characteristics [7]. The increasing frequency, duration, and intensity of droughts underscore the importance of effective monitoring. Among the various indexes used to assess the severity of these phenomena, the streamflow drought index (SDI) is widely recognized as a simple yet effective method for evaluating hydrological droughts. Numerous studies conducted in regions such as Northern Europe [8], Australia [9], Ethiopia [10], India [11], and Turkey [12] have demonstrated the severity of droughts, highlighting the need for robust mitigation measures and a deeper understanding of the relationship between drought and climate change [13,14]. Furthermore, the Intergovernmental Panel on Climate Change (IPCC) and the Coupled Model Intercomparison Project (CMIP) have been instrumental in developing models to predict future climate changes, with CMIP5 showing enhanced performance in simulating global precipitation trends [15].

Northeastern Thailand is significantly affected by climate change, leading to decreased rainfall and streamflow, with projections indicating a 13–19% reduction in annual streamflow and shifts in seasonal patterns [16]. This decline has severely impacted the agricultural sector, especially rice farming, where yields are expected to decrease due to higher temperatures and altered rainfall patterns [15]. The Lam Chiang Kri Watershed (LCKW) is particularly vulnerable due to its geographical limitations and sandy soil, whose water-holding capacity is poor. The region experienced a 24.52% decrease in rice production during strong El Niño events, further illustrating the severe impact of drought [17]. Additionally, the region's climate variability, frequent droughts, and issues such as soil erosion and salinity exacerbate existing agricultural challenges. Consequently, sustainable water resource management and improved agricultural practices are essential to addressing these issues [18–20]. This study presents a novel perspective on the hydrological and climatic characteristics of Northeastern Thailand, a region whose unique features have been largely underexplored in existing literature. By integrating hydrological drought analysis with return period assessments under climate change scenarios, the research addresses a significant knowledge gap [21]. By focusing on Northeastern Thailand's unique challenges, the study provides valuable insights into the region's vulnerability to future droughts and contributes to the development of effective water resource management and planning strategies.

In this study, we aimed to fill the existing research gap by evaluating the impact of climate change on streamflow in the Lam Chiang Kri Watershed, assessing hydrological drought by using the streamflow drought index (SDI), and characterizing this phenomenon across different return periods in two climate change scenarios. We utilized the SWAT model, a well-established tool for simulating streamflow, based on downscaled climate projections from selected global climate models (GCMs) in two emission scenarios: RCP4.5 and RCP8.5. The analysis focuses on projections for the years 2029 and 2039, incorporating observational streamflow data for the reference period to estimate future streamflow and calculate hydrological drought.

## 2. Materials and Methods

### 2.1. Study Area

The Lam Chiang Kri Watershed (LCKW) is located in northeastern Thailand, within the Isan Plateau, and serves as the upper branch of the Mun Watershed. Covering 2959.59 square kilometers, it has an elevation range of 145 m to 593 m above sea level (Figure 1) [22]. The terrain slopes from west to east, with predominantly laterite soil, which is a type of sandy loam with poor water-holding capabilities. As a result, hydrographs in this region display sharp rising limbs, high peaks, and steep recess limbs, indicating that rainfall quickly runs off rather than soaking into the ground, leading to rapid changes in streamflow levels [23].

The LCKW experiences lower rainfall and higher temperatures than other regions in Thailand. According to data from the Thai Meteorological Department (TMD), the area receives an average annual rainfall of 947.66 mm and has an average temperature of 33 °C. The LCKW has distinct wet and dry seasons driven by monsoons [24]. The southwestern monsoon brings heavy rain from mid-May to mid-October, while the northeastern monsoon causes the dry season from mid-October to mid-February, with a transitional period occurring from mid-February to mid-May. Additionally, the Roy-al Irrigation Department of Thailand (RID) reports an average annual streamflow of 2661.51 m3/s, with 91.78% occurring during the rainy season and the remaining 8.22% in the dry season. Furthermore, an analysis by Thailand's Land Development Department (LDD) indicated that 88.89% of the land use in the LCKW is agricultural, yet only 22.09% is irrigated. This highlights the area's high vulnerability to drought, with 45.89% of land being classified as high risk, 29.72% as moderate risk, and 24.39% as low risk. These factors underscore the region's susceptibility to drought, which significantly affects agriculture and water resources.

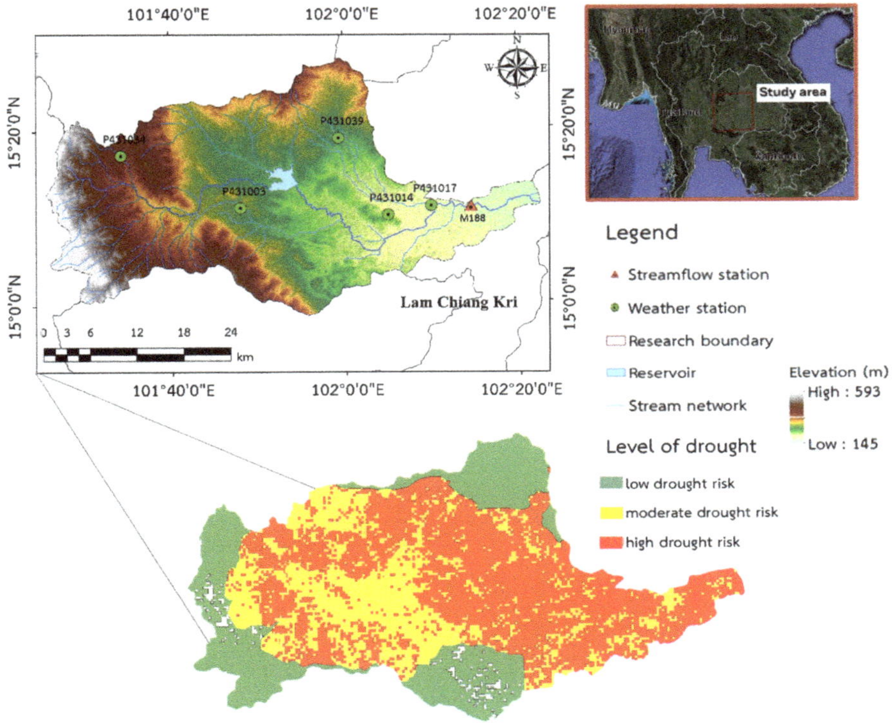


**Figure 1.** The geographical location, topographical features, historical drought patterns, and weather stations of the study area.

### 2.2. Data Collection

#### 2.2.1. Meteorological Data

Daily meteorological data, including precipitation, temperature, humidity, wind speed, and radiation, were obtained from the Thai Meteorological Department for the period from 1992 to 2022 [24]. Within the Lam Chiang Kri Watershed, data were collected from five meteorological stations: P431003 (Dan Khun Thot), P431014 (Non-Thai), P431017 (Non-Sung), P431034 (Theparak), and P431039 (Phra Thong Kham). We applied the Thiessen method to ensure an accurate representation of rainfall across the watershed. This spatial interpolation technique divides the area into polygons, with each polygon assigned the rainfall data from the nearest meteorological station. As a result, it provides a precise and reliable distribution of rainfall data across the region [25]. By doing so, the method provides a more representative average of rainfall distribution across the watershed. The processed data were then used as inputs for bias correction in climate models to evaluate past, present, and future climate scenarios.

#### 2.2.2. Hydrological Data

Daily streamflow data in this study were obtained from the Irrigation Hydrology Center, Royal Irrigation Department Thailand [26]. This station is M188 (Ban Bua). The data covers April 2010–March 2021.

#### 2.2.3. Topographic, Soil, and Land Use Data

The digital elevation model (DEM) was used in this study to represent the topographic condition of the study area. This DEM has a resolution of 12.5 m. These data were retrieved from the National Aeronautics and Space Administration (NASA) [27].

This study used a 2021 land use map and soil map of the LCKW, created by Thailand's Land Development Department (LDD), with reference to a spatial resolution suitable for detailed watershed analysis.

### 2.3. Methodology

In this study, we utilized daily rainfall data from five gauges of the Thai Meteorological Department (TMD), covering the historical period from 1992 to 2022. Additionally, daily streamflow data from the gauge at the M188 station, provided by the Royal Irrigation Department of Thailand (RID), were used for the historical period from 2010 to 2021. The locations of these gauges are illustrated in Figure 1, while a schematic diagram of the overall framework is shown in Figure 2. For the projected periods, two specific years were selected, the 5th year (2029) and the 15th year (2039) from the current year, to assess the impact of climate change on hydrological drought. Observational rainfall and streamflow data were crucial to calibrating and validating the Soil and Water Assessment Tool (SWAT) model version 2012, as well as calculating the baseline (2010–2021) hydrological drought index. After performing bias correction, the ability of the global climate models (GCMs) to generate streamflow for the baseline period was evaluated. The output from all selected GCMs was then analyzed, with particular emphasis on the drought index results obtained from the best-performing GCMs.

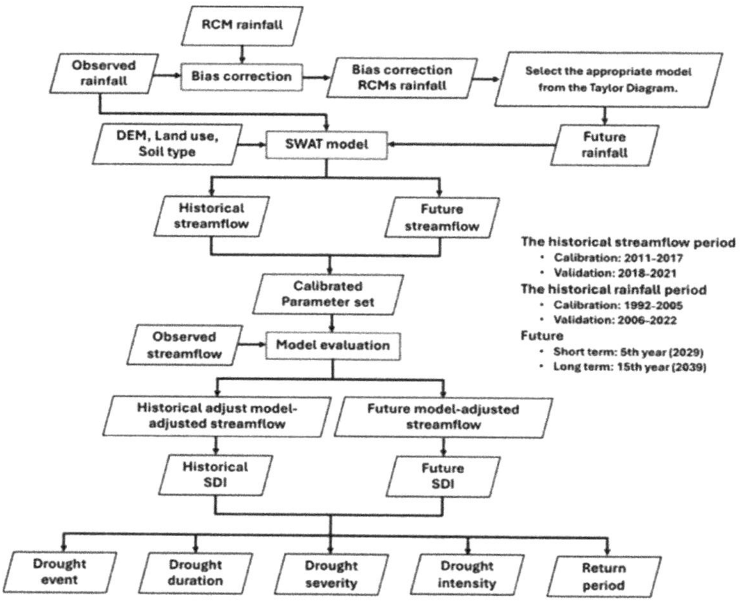


**Figure 2.** Overview of global climate model (GCM) methodology for hydrological drought assessment.

2.3.1. Global Climate Models (GCMs) and Climate Scenarios

In this study, we employed three well-regarded global climate models (GCMs) from the CMIP5 dataset: EC-Earth3, HadGEM2, and MPI-ESM-MR [28,29]. These models were selected for their strong capabilities in simulating historical climate variability and accurately projecting future scenarios. EC-Earth3 is particularly effective in high-resolution assessments of extreme events like floods, droughts, and heatwaves. HadGEM2 is known for its precision in modeling global warming and extreme weather conditions, making it ideal for hydrology and drought studies. MPI-ESM-MR excels in simulating complex climate interactions, such as monsoon dynamics, and is especially useful for assessing long-term drought risks in Southeast Asia [30–32]. These models were chosen for their proven ability to represent precipitation, which is a crucial variable in hydrological responses and potential future hydrological drought scenarios. These scenarios include projected frequency, duration, and severity of droughts as influenced by different climate change pathways. Specifically, CMIP5 was chosen for its ability to address the diverse and complex geography of Southeast Asia, which includes mountain ranges, major rivers, and coastal regions highly vulnerable to climate change. Its validation in numerous regional studies ensures its suitability for addressing local complexities [28]. To analyze future climate impacts, we used precipitation data from these models under two Representative Concentration Pathways (RCPs), which are greenhouse gas concentration trajectories adopted by the Intergovernmental Panel on Climate Change (IPCC). Specifically, RCP 4.5 represents a moderate climate impact scenario where radiative forcing stabilizes at 4.5 $W/m^2$ by 2100 [33], while RCP 8.5 depicts a high-emission scenario with radiative forcing reaching 8.5 $W/m^2$ by 2100, while RCP 8.5 depicts a high-emission scenario with radiative forcing reaching 8.5 $W/m^2$ by 2100. The comprehensive datasets and extensive validation of these models in similar climates ensure the reliability and robustness of our projections.

To ensure the accuracy of the projected climate data, we refined the GCM outputs through dynamic downscaling [34], improving alignment with local climate patterns and enhancing their suitability for regional impact assessments. The downscaled models were validated by comparing their outputs with observed climate data from 1992 to 2022, using statistical metrics such as correlation coefficient (r), root mean square error (RMSE), and standard deviation (SD). The models' performance was visually summarized using a Taylor diagram, allowing for comparative evaluation [35]. For our analysis of future climate impacts, we focused on two specific years, 2029 and 2039, representing the 5th and 15th years within the projected timeframe. These years were chosen to capture both near-term and mid-term climate impacts, offering insights into potential changes in hydrological patterns, particularly regarding drought conditions under different RCP scenarios. To further ensure the accuracy of these projections, we compared the models' rainfall predictions for these years with observed rainfall data from 2004 to 2022, allowing us to assess the models' reliability in projecting future climatic conditions.

2.3.2. SWAT Model

(1) Model Description

The SWAT model, a semi-distributed, process-based hydrological tool developed by the United States Department of Agriculture (USDA), was employed in this study to simulate watershed processes [36], with a particular focus on assessing the impacts of climate change. The model utilized a comprehensive set of input data, including climate variables sourced from the Thai Meteorological Department, topographical data derived from a 12.5-m resolution DEM, soil characteristics provided by the Land Development Department of Thailand, and land use data reflecting both current and historical patterns. The construction of the model involved delineating the watershed, creating hydrologic response units (HRUs) based on land use, soil type, and slope, and integrating these datasets to accurately represent the hydrological processes within the watershed. Calibration and validation of the model were conducted using multi-temporal observed streamflow data from 2010 to 2021, ensuring the model's precision and reliability in simulating the watershed's behavior under various climate scenarios [37].

In this study, the SWAT model was also utilized to delineate the watershed, divide it into sub-watersheds, and create HRUs based on land use, soil type, and slope data. The Land Development Department of Thailand classified land use into twelve primary categories, while the soil map identified forty-four distinct soil types within the study area. Additionally, the slope was categorized into five classes: flat (0–2%), sloping (2–5%), hilly (5–15%), steep (15–35%), and very steep (>35%).

A water balance equation was the basis for the SWAT model, represented as follows (Equation (1)):

$$SW_t = SW_0 + \sum\left(R_{day} - Q_{surf} - E_a - W_{seed} - Q_{gw}\right) \tag{1}$$

where $SW_0$ and $SW_t$ (mm) are the initial and final soil water on a given day, and $R_{day}$, $Q_{surf}$, $E_a$, $W_{seed}$, and $Q_{gw}$ (mm) are the rainfall, runoff, ET, water seepage to the upper soil layer, and return flow on that day, respectively.

The SWAT model used the Soil Conservation Service curve number (SCS-CN) approach to compute surface runoff in the study area. The SCS-CN equation is shown by Equation (2), as follows:

$$Q_{surf} = \frac{\left(R_{day} - I_a\right)^2}{\left(R_{day} - I_a + S\right)} \tag{2}$$

where $Q_{surf}$ is daily surface runoff (mm); $R_{day}$ is daily rainfall depth (mm); $I_a$ is the initial abstraction (mm); and $S$ is the retention parameter (mm). The retention parameter

$S$ is not fixed and can be affected by factors such as slope, soil, and land-use management. Mathematically, the retention parameter can be represented as Equation (3), as follows:

$$S = 254 \times \left(\frac{100}{CN} - 1\right) \tag{3}$$

where $S$ is the retention parameter (mm), and $CN$ is the curve number. The curve number ranges from 0 to 100, with 100 indicating no potential retention and 0 reflecting infinite potential retention [38].

(2) Model Setup

The SWAT model calibration and validation process requires careful consideration of observation streamflow data from the M188 station. The data are divided into 80% for calibration and 20% for validation, with periods selected from 2010 to 2021. The simulation runs for 12 years, starting from 1 January 2010, to 31 December 2021. Nine sub-watersheds were created in the study area, with a threshold of 10% for land use, 10% for soil, and 10% for slope, resulting in 108 hydrologic response units.

(3) Model Evaluation

The study used SWAT-CUP software version 5.1.6 with the Sequential Uncertainty Fitting (SUFI) algorithm to calibrate a model [39], which can handle a large number of parameters and combine sensitive analysis and improvement [37].

The model's performance was compared using three statistical performance indices: Nash and Sutcliffe Efficiency (NSE) following Equation (4) [40]; the coefficient of determination ($R^2$), following Equation (5) [41]; percent bias (PBIAS), following Equation (6) [42]; and Kling–Gupta Efficiency (KGE), following Equation (7) [43], to evaluate its daily stream-flow performance during calibration and validation phases.

$$NSE = 1 - \frac{\sum_{i=1}^{n}\left(Q_{obs} - Q_{sim}\right)^2}{\sum_{i=1}^{n}\left(Q_{obs} - \overline{Q}_{obs}\right)^2} \tag{4}$$

$$R^2 = \left[ \frac{\sum_{i=1}^{n} \left[ \left( Q_{obs} - \overline{Q}_{obs} \right) \left( Q_{sim} - \overline{Q}_{sim} \right) \right]}{\left[ \sum_{i=1}^{n} \left( Q_{obs} - \overline{Q}_{obs} \right)^2 \right]^{0.5} \left[ \sum_{i-1}^{n} \left( Q_{sim} - \overline{Q}_{sim} \right)^2 \right]^{0.5}} \right]^2 \tag{5}$$

$$PBIAS = \frac{\sum_{i=1}^{n} (Q_{obs} - Q_{sim})}{Q_{obs}} \times 100 \tag{6}$$

$$KGE = 1 - \sqrt{(r-1)^2 + (\alpha - 1)^2 + (\beta - 1)^2} \tag{7}$$

$Q_{obs}$ and $Q_{sim}$ represent observed and simulated values, respectively. The NSE value of the model should be more than 0.50, while the $R^2$ should be at least 0.7. PBIAS should not exceed 25 percent to be acceptable. KGE can be categorized as good (KGE $\geq$ 0.75), intermediate (0.75 > KGE $\geq$ 0.5), poor (0.5 > KGE > 0), and very poor (KGE $\leq$ 0).

### 2.3.3. Hydrological Drought Index

The streamflow drought index (SDI) is a key tool for assessing hydrological drought severity, where positive values indicate wetter conditions, and negative values signal the presence of drought. Calculated using monthly streamflow data, the SDI aids in managing drought and water scarcity across various time frames, encompassing both dry and wet seasons [44]. Specifically, the SDI-3, which tracks drought over a 3-month period, is particularly valuable for monitoring agricultural droughts and their impact on crops, whereas the SDI-6, calculated over 6 months, offers deeper insights into hydrological droughts that affect both surface and groundwater resources [45]. By analyzing both SDI-3 and SDI-6 together, a more comprehensive understanding of drought conditions can be achieved across both short-term and long-term time scales. Moreover, these time scale choices can be further refined depending on the specific climate and water resource focus of the study area, as detailed in Equation (8) [44].

$$SDI = \frac{V_{n,q} - V_{qm}}{S_q} \tag{8}$$

where $V_{n,q}$ represents the cumulative streamflow volume for period ($n$) and quarter ($q$) while $V_{qm}$ and $S_q$ are the mean and the standard deviation of cumulative streamflow volumes of the reference period, respectively. The classification of hydrological drought based on the SDI (Table 1) offers a detailed understanding of drought characteristics.

**Table 1.** Classification of hydrological drought based on SDI.

| State | Description | Range |
|---|---|---|
| 1 | No drought | $0 \leq$ SDI |
| 2 | Mild drought | $-1 \leq$ SDI < 0 |
| 3 | Moderate drought | $-1.5 \leq$ SDI < $-1$ |
| 4 | Severe drought | $-2 \leq$ SDI < $-1.5$ |
| 5 | Extreme drought | SDI $\leq -2$ |

The Theory of Runs (ToR) is a statistical method used to analyze drought characteristics [46], including drought event (DE), drought duration (DD), drought severity (DS), and drought intensity (DI) (Figure 3). DE is identified when the SDI value falls below a critical threshold. DD represents the duration of drought in months with negative SDI values, while DS is the sum of the absolute values of the SDI during a DE. DI can be defined as the absolute lowest value of the index (DI1) or the ratio of DS to DD in a DE (DI2).

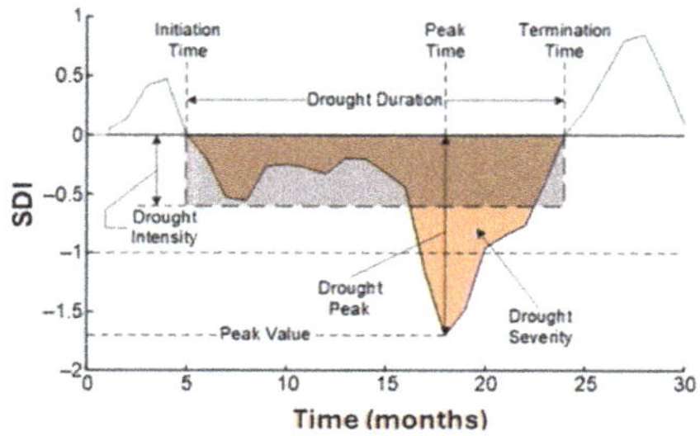


**Figure 3.** A Theory of Runs illustration of a drought event and the drought indicators [47].

2.3.4. Scenarios Analysis in Different Return Periods

The CumFreq software version 5.0. (https://www.waterlog.info/cumfreq.htm; accessed on 28 January 2024) was employed to determine the most suitable statistical distribution for characterizing drought events across various return periods (5, 10, 25, and 50 years) and time scales (3 and 6 months). CumFreq utilizes multiple probability distribution functions to analyze the input data and subsequently recommend the most appropriate distribution for drought characterization [48]. The absolute values of the streamflow drought index (SDI) were input into CumFreq to derive these distributions.

To further analyze the impact of droughts over different periods, wavelet analysis was applied to assess the variability of SDI values at different time scales and across varying return periods. This method provided deeper insights into the temporal patterns and severity of droughts under different climate scenarios. Additionally, a geostatistical approach was employed to interpolate streamflow and SDI values, which were then visualized as contour maps. These maps were generated using accurate variogram models, which are crucial for interpreting the spatial distribution of natural phenomena like drought [49]. Moreover, the relationship between streamflow and absolute SDI was explored using Surfer 21.1.158 software, enabling the creation of three-dimensional diagrams with contour lines based on the Kriging interpolation method [50].

## 3. Results

The investigation of critical hydrological droughts in the Lam Chiang Kri Watershed (LCKW) in CMIP5 climate change scenarios was divided into three key areas, i.e., the calibration and validation of the SWAT model, identification of historical drought characteristics, and assessment of climate change impacts on hydrological drought, as detailed in the following sections.

### *3.1. Calibration and Validation of SWAT Model*

The SWAT model simulation for the period from 2010 to 2021, supported by data from a hydrological station within the Lam Chiang Kri Watershed, allowed us to effectively analyze the watershed's hydrological responses to varying meteorological conditions. The calibration and validation performed by using the SUFI-2 algorithm within SWAT-CUP ensured the accuracy of the streamflow patterns, significantly enhancing the reliability of the results (Table 2).

**Table 2.** Sensitivity parameters in the SWAT-CUP model of LCKW.

| Parameter | *t*-Stat | *p*-Value | Fit Value | Min Value | Max Value |
|---|---|---|---|---|---|
| 1: R__CN2.mgt | −5.48 | 0.01 | 40.395 | 35 | 100 |
| 2: R__SOL_AWC(..).sol | −2.84 | 0.04 | 0.343 | −0.2 | 0.4 |
| 3: R__ESCO.hru | 2.16 | 0.08 | 0.21525 | 0.1 | 0.35 |
| 4: V__GW_DELAY.gw | 1.82 | 0.14 | 155.5 | 0 | 500 |
| 5: R__SLSUBBSN.hru | −1.71 | 0.17 | 56.90 | 50 | 150 |

A sensitivity analysis identified five critical parameters—CN2, ESCO, SOL_AWC, GW_DELAY, and SLSUBBSN—highlighting their significant influence on the streamflow simulation. These findings provide a robust foundation for future water resource management and drought mitigation strategies in the LCKW.

The calibration and validation phases in SWAT model development depend on accurate observational streamflow data. For this study, data from the M188 station, covering the period from 2010 to 2021, were utilized. Figure 4 shows that the calibrated SWAT model for the M188 station closely matched the observational data patterns.

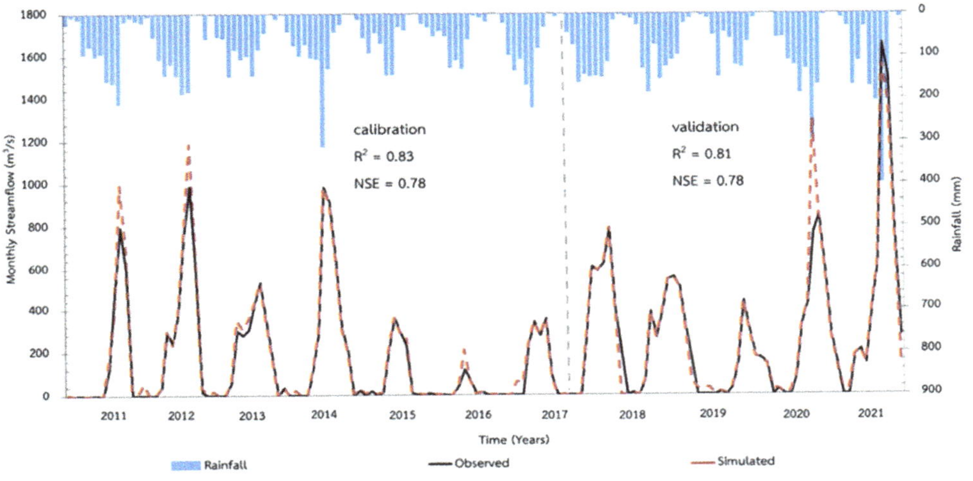


**Figure 4.** The monthly simulated and observed streamflow comparison for the M188 station during the calibration (2010–2017) and validation (2018–2021) periods. The two periods are separated by vertical dashed lines within the graphs.

For statistical evaluation, $R^2$, NSE, PBIAS, and KGE values were used. Throughout the calibration and validation periods, the streamflow station data exhibited $R^2$ and NSE values above 0.75, indicating good to very good performance. The PBIAS values were maintained below 25%, aligning with the preferred threshold. The KGE values were classified in the intermediate category, as detailed in Table 3.

**Table 3.** Statistical parameters of the SWAT model based on SWAT-CUP.

| Statistic Parameters | Calibration (2011–2017) | Validation (2018–2021) | Total (2011–2021) |
|---|---|---|---|
| $R^2$ | 0.83 | 0.81 | 0.82 |
| NSE | 0.78 | 0.78 | 0.78 |
| PBIAS | 12.0 | 28.04 | 20.02 |
| KGE | 0.64 | 0.46 | 0.55 |

### 3.2. *Identification of Historical Drought Characteristics*

The historical drought characteristics were calculated using the streamflow drought index (SDI) and divided into two periods: SDI-3 and SDI-6. The figure illustrates the SDI variations in the LCKW from 2010 to 2021, focusing on these two different accumulation periods. According to the 3-month accumulation period (SDI-3) analysis, the graph indicates frequent fluctuations in the SDI values (Figure 5), with significant drought periods around 2014–2015, where the mean SDI-3 value dropped to approximately −1.75. During 2020–2021, the SDI-3 values decreased to levels as low as −2.25, suggesting that short-term drought events are common and often occur annually. The 6-month accumulation period

(SDI-6) analysis showed less frequent but more severe and prolonged droughts (Figure 5). Notably, in 2015, the SDI-6 value reached approximately −2.5, and in 2020, it dropped further, to around −2.74, indicating that medium-term droughts, while less frequent, tend to be more intense and prolonged. Both the SDI-3 and SDI-6 analyses indicated that severe and extreme drought events occurred almost every year, highlighting the persistent and recurring nature of drought conditions in the region.

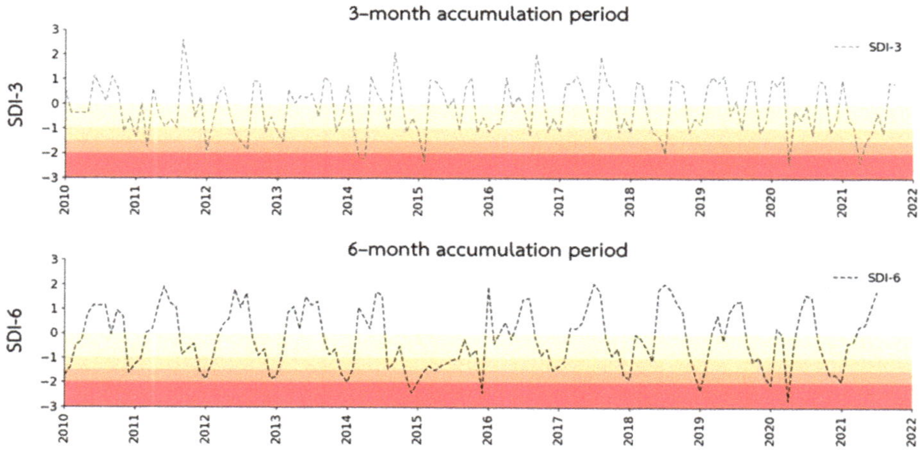


**Figure 5.** The temporal variation in the spatial averaged time series of the SDI in the LCKW at the 3- and 6-month time scales calculated based on the period of 2010–2021. The color scale from yellow to red represents mild to moderate, severe, and extreme drought categories, respectively.

The analysis indicated an average of 2.67 drought events per year, with a maximum duration of 3 months and a peak severity of −31.97. The SDI-6 analysis showed less frequent yet more severe and prolonged droughts, notably in 2015 and 2020, with an average of 1.25 drought events per year, a maximum duration of 6 months, and a peak severity of −43.04. Both indexes underscore the persistent and recurrent nature of drought conditions, highlighting the necessity for effective water resource management strategies to mitigate both short-term and medium-term drought risk. The maximum intensity values for SDI-3 were −2.44 in 2015 (DI1) and −1.35 (DI2), whereas for SDI-6, they were −2.74 (DI1) in 2020 and −1.69 (DI2) (Table 4).

**Table 4.** The historical drought characteristics in the LCKW represented by the SDIs for 3- and 6-month accumulation periods.

| | Hydrological Drought | |
|---|---|---|
| | **SDI-3** | **SDI-6** |
| Average drought event (time/year) | 2.67 | 1.25 |
| Total number of drought events (times) | 32 | 15 |
| Maximum drought duration (months) | 23 | 36 |
| Maximum drought severity | −31.97 | −43.04 |
| Maximum drought intensity based on DI1 | −2.44 | −2.74 |
| Maximum drought intensity based on DI2 | −1.35 | −1.69 |

### 3.3. *Assessment of Climate Change Impacts on Hydrological Drought*

#### 3.3.1. The Selection of the Fittest GCM

In this study, we utilized daily rainfall observations from five weather stations located within the Lam Chiang Kri Watershed (LCKW) (as shown in Figure 1). The rainfall data were aggregated and analyzed by using the Thiessen method to accurately represent the spatial distribution of rainfall across the watershed. The observational data covered the historical period from 1992 to 2022, which was divided into two distinct phases: 1992–2005 for calibration and 2006–2022 for validation. These periods were used to assess the performance of three selected CMIP5 GCMs, i.e., EC-Earth3, HadGEM2, and MPI-ESM-MR, in the RCP4.5 and RCP8.5 scenarios.

The data were processed to ensure consistency and accuracy before being used in Taylor diagram analysis. To create the Taylor diagram, data for 1992–2022 were used to assess how accurately the models predicted results compared with the observational data during that period, which were employed to assess the fit of these models. Among the models, the EC-Earth3 model (red dot) demonstrated the highest correlation with the observational data, with a correlation coefficient of 0.65, a standard deviation (SD) of 1.98, and a root mean square error (RMSE) of 3.41, making it the most accurate model. The HadGEM2 model (blue dot) also showed a strong correlation but had a higher RMSE. The MPI-ESM-MR model (green dot) exhibited moderate correlation and RMSE values (Figure 6).

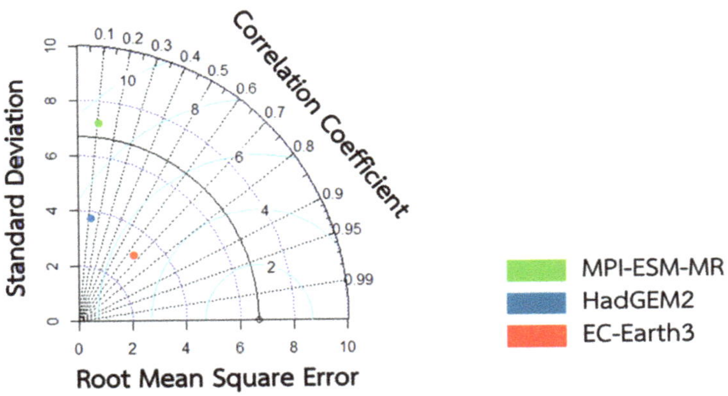


**Figure 6.** The Taylor diagram illustrates the suitability of different climate models for projecting rainfall in the LCKW.

### 3.3.2. Future Rainfall

The future rainfall analysis compared monthly averages from the observational periods with projections in the RCP4.5 and RCP8.5 scenarios. Figure 7 presents the EC-Earth3 model's average monthly rainfall predictions for 2029 and 2039, alongside the baseline data for 2004–2022 (Table 5), to evaluate changes in both climate scenarios.

In 2029, which represents the 5th year from the baseline reference year (2024), in the RCP4.5 scenario, the projected annual rainfall totals 1311.94 mm, with the highest monthly rainfall value occurring in September (383.03 mm) and the lowest one in January (11.46 mm). This represents a predicted increase of 38.44% in future rainfall compared with the baseline. In the RCP8.5 scenario, the projected annual rainfall is 892.97 mm, with the highest value in August (215.53 mm) and the lowest one in February (11.90 mm), indicating a decrease of 5.77% compared with the baseline.

In 2039, which represents the 15th year from the baseline reference year (2024), in the RCP4.5 scenario, the projected annual rainfall is 1101.06 mm, with the highest monthly rainfall value in August (248.36 mm) and the lowest one in January (14.80 mm), marking a 16.19% increase compared with the baseline. In the RCP8.5 scenario, the projected annual rainfall totals 805.46 mm, with the highest value in October (218.96 mm) and the lowest one in January (1.33 mm), reflecting a 15.00% decrease compared with the baseline (Table 5).

The findings indicate that there are considerable differences in the variability of monthly rainfall between the two future periods, with more substantial changes being predicted for 2039, particularly in the RCP8.5 scenario, where the most significant rainfall declines are anticipated.

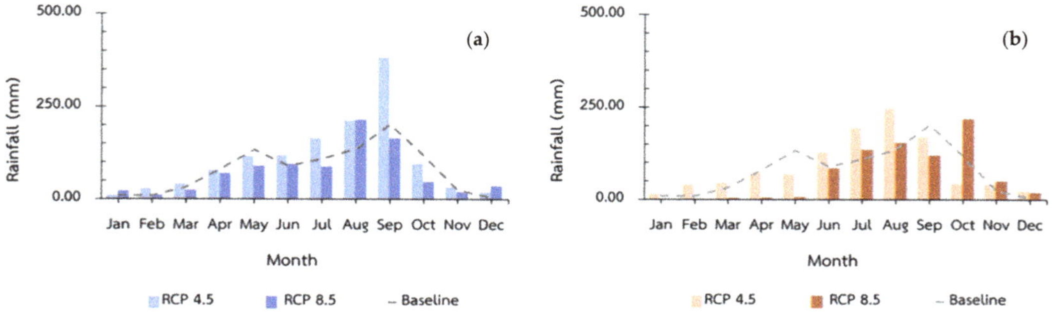


**Figure 7.** A comparison of streamflow between the baseline period (2004–2022) and the projections for 2029 (**a**) and 2039 (**b**) in the RCP4.5 and RCP8.5 future climate scenarios.

**Table 5.** Future rainfall changes in the RCP4.5 and RCP8.5 scenarios for 2029 and 2039 compared with the baseline period of 2004–2022.

| Time | Baseline (2004–2022) | RCP4.5 | | RCP8.5 | |
|---|---|---|---|---|---|
| | (mm) | (mm) | % Change | (mm) | % Change |
| 2029 | 947.64 | 1311.94 | 38.44 | 892.97 | −5.77 |
| 2039 | 947.64 | 1101.06 | 16.19 | 805.46 | −15.00 |

### 3.3.3. Future Streamflow

In this study, we analyzed variations in streamflow by comparing historical data with model simulations, focusing on mean annual streamflow in the RCP4.5 and RCP8.5 climate scenarios. Figure 8 presents the predicted average monthly streamflow for 2029 and 2039 compared with the baseline period of 2010–2021 (Table 6) to evaluate potential changes in each scenario.

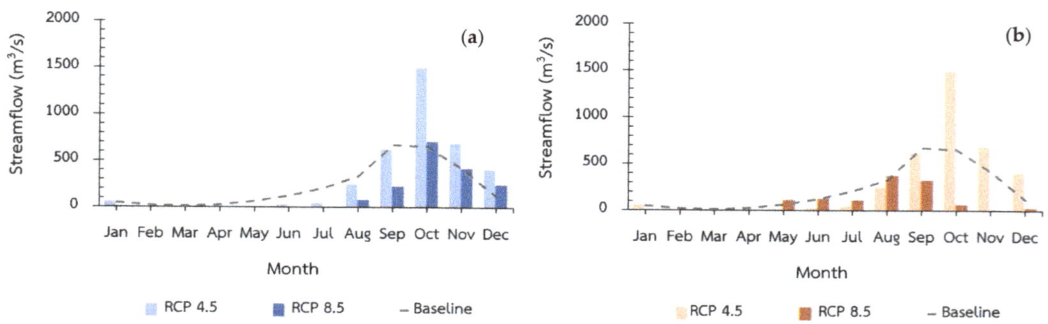


**Figure 8.** A comparison of streamflow between the baseline period (2010–2021) and the projections for 2029 (**a**) and 2039 (**b**) in the RCP4.5 and RCP8.5 future climate scenarios.

**Table 6.** A comparison of the amount of streamflow between the baseline (2010–2021) and the RCP4.5 and RCP8.5 future climate scenarios for both 2029 and 2039.

| Months | Baseline ($m^3$/s) | | Future Streamflow ($m^3$/s) | | | |
|---|---|---|---|---|---|---|
| | | | RCP4.5 | | RCP8.5 | |
| | | | 2029 | 2039 | 2029 | 2039 |
| | Observation | Simulation | Simulation | Simulation | Simulation | Simulation |
| January | 45.96 | 46.59 | 60.83 | 45.78 | 4.94 | 4.94 |
| February | 19.30 | 19.30 | 1.20 | 10.92 | 0.00 | 5.76 |
| March | 9.77 | 10.66 | 0.13 | 9.66 | 0.00 | 6.69 |
| April | 24.54 | 25.14 | 0.13 | 9.66 | 0.19 | 0.19 |
| May | 62.52 | 63.23 | 12.08 | 6.08 | 0.07 | 116.05 |
| June | 122.29 | 123.02 | 30.30 | 7.45 | 0.33 | 128.14 |
| July | 205.78 | 206.47 | 46.95 | 15.21 | 4.57 | 114.43 |
| August | 329.76 | 330.48 | 249.36 | 159.97 | 80.37 | 378.82 |
| September | 626.43 | 671.42 | 624.79 | 474.93 | 225.28 | 325.10 |
| October | 652.15 | 653.03 | 1495.52 | 1495.52 | 708.20 | 65.79 |
| November | 419.28 | 419.99 | 691.80 | 593.66 | 421.28 | 10.70 |
| December | 143.72 | 119.69 | 410.01 | 363.35 | 247.15 | 30.93 |
| Total runoff | 2661.51 | 2689.02 | 3623.09 | 3192.18 | 1692.38 | 1187.54 |
| Wet period ($m^3$/s) | 2442.76 | 2492.78 | 3150.93 | 2762.47 | 1440.29 | 1139.21 |
| Dry period ($m^3$/s) | 218.74 | 196.24 | 472.17 | 429.71 | 252.09 | 48.33 |
| (%) Percentage change | - | - | 34.74 | 18.71 | −37.06 | −55.84 |

In 2029, in the RCP4.5 scenario, the annual streamflow is projected to be 3623.09 $m^3$/s, with the highest flow value occurring in October (1495.52 $m^3$/s) and the lowest one in February and March (0.13 $m^3$/s). The monsoon season is expected to contribute 86.97% (3150.93 $m^3$/s) of the total streamflow, with the dry season contributing 13.03% (472.17 $m^3$/s), indicating a 34.74% increase compared with the baseline period. In the RCP8.5 scenario, the streamflow is projected to decrease to 1692.38 $m^3$/s, with the highest flow in October (708.20 $m^3$/s) and no flow in February and March. The monsoon season is expected to contribute 85.10% (1440.29 $m^3$/s) of the total streamflow, with the dry season contributing 14.90% (252.09 $m^3$/s), representing a 37.06% decrease compared with the baseline period (Table 6).

In 2039, in the RCP4.5 scenario, the projected streamflow is 3192.18 $m^3$/s, reflecting an 18.71% increase compared with the baseline. October again has the highest flow (1495.52 $m^3$/s), and the lowest flows are expected in March and April (9.66 $m^3$/s). The monsoon season is projected to contribute 86.54% (2762.47 $m^3$/s) of the total streamflow, with the dry season contributing 13.46% (429.71 $m^3$/s). Conversely, in the RCP8.5 scenario for 2039, streamflow is projected to further decrease to 1187.54 $m^3$/s, a 55.84% decrease compared with the baseline. The peak flow is expected in August (378.82 $m^3$/s) and the lowest value in April (0.19 $m^3$/s). The monsoon season is expected to contribute 95.93% (1139 $m^3$/s) of the total, with the dry season accounting for only 4.06% (48.33 $m^3$/s).

The results highlight significant disparities in the variability in monthly streamflow between the two future time frames, with more pronounced changes being observed in 2039, especially in the RCP8.5 scenario, where the steepest declines in streamflow are projected. These changes in streamflow mirror the patterns seen in rainfall variability, indicating a strong correlation between precipitation and streamflow responses to climate change.

### 3.3.4. Future Hydrological Drought Characteristics

The future hydrological drought characteristics, as projected based on the SDI for the years 2029 and 2039, were analyzed in two climate scenarios, RCP4.5 and RCP8.5, based on 3-month (SDI-3) and 6-month (SDI-6) accumulation periods (Figure 9), as detailed below.

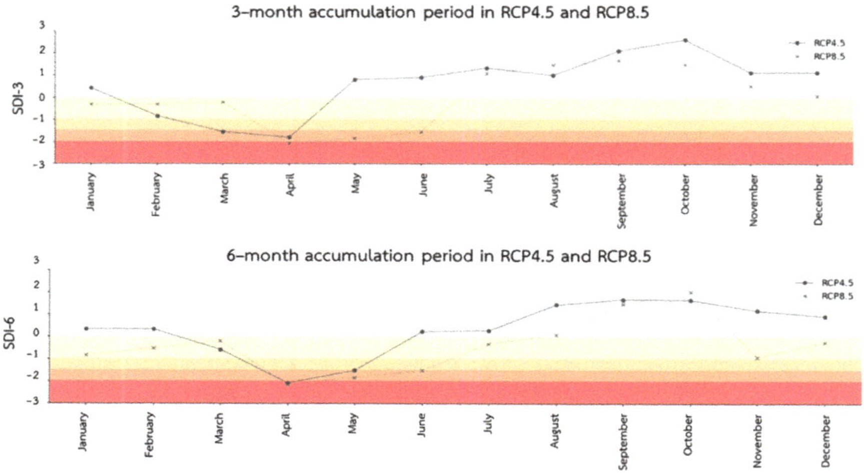


(**a**)

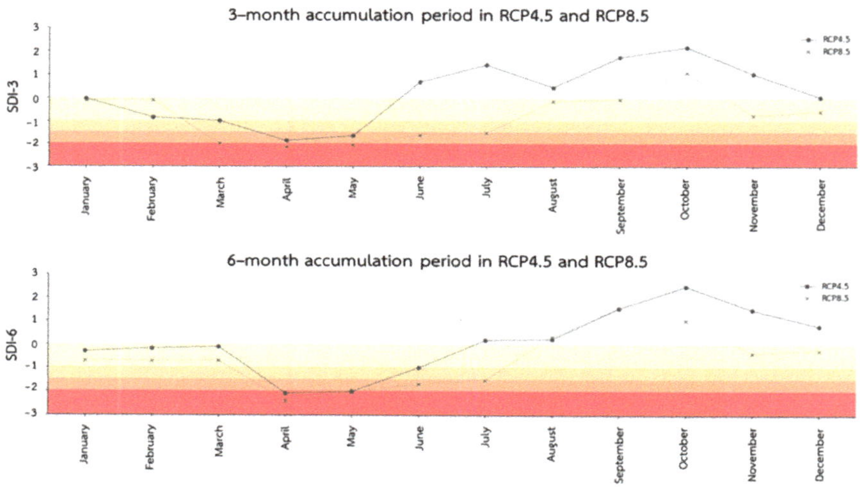


(**b**)

**Figure 9.** The temporal variation in the spatial averaged time series of the SDI in the LCKW at 3- and 6-month time scales calculated based on the EC-Earth3 model: 2029 (**a**); 2039 (**b**). The color scale from yellow to red represents mild to moderate, severe, and extreme drought categories, respectively.

The analysis shows that in 2029, the SDI-3 values in RCP4.5 indicate drier conditions from January to April, with improvements from May to September and a decline towards the year's end. The RCP8.5 scenario follows a similar pattern but with more severe droughts. In terms of the SDI-6 values, both scenarios exhibit dry conditions early in the year, with some recovery from April to July and a decline towards the year's end, with RCP8.5 showing more intense droughts. For 2039, both SDI-3 and SDI-6 projections in RCP4.5 and RCP8.5 suggest worsening droughts, particularly in RCP8.5, with severe droughts expected from April to July. The projections highlight an increasing trend in drought severity and frequency, especially in the higher-emission scenario (RCP8.5), emphasizing the urgent need for effective water management and climate adaptation measures to mitigate the adverse impacts of these projected changes.

3.3.5. Analysis of SDI in Different Return Periods

We further investigated the relationship between streamflow and the absolute SDIs across different return periods and time scales by using wavelet analysis (Figure 10).

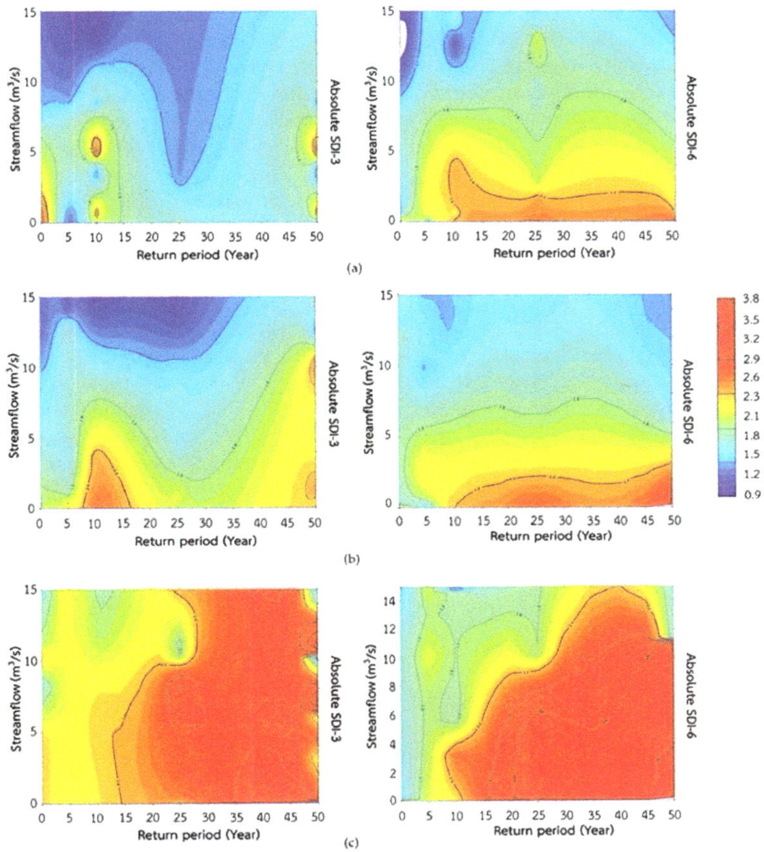


**Figure 10.** The wavelet analysis results showing the relationship among streamflow ($m^3/s$), absolute SDIs, and return periods for the baseline (**a**), RCP4.5 (**b**), and RCP8.5 (**c**). The horizontal axis represents return periods (years), and the vertical axis represents average streamflow ($m^3/s$). Bold blue lines indicate areas of low drought severity, while bold red lines mark areas of high severity. Contour lines highlight transitions between severity levels, with the color gradient further illustrating drought severity (blue for low and red for high).

For SDI-3, return periods of 2–10 years exhibit the highest drought severity, with SDI values ranging from 2.08 to 2.86, indicating that the most severe droughts occur over shorter periods. In contrast, longer return periods (over 10 years) show less intense and shorter droughts. In the RCP4.5 scenario, the most severe droughts are observed within 2–10 year return periods, rather than increasing with longer return periods. Conversely, the RCP8.5 scenario shows more severe and widespread droughts, particularly in return periods of 25–50 years, with absolute SDI values ranging from 2.26 to 4.34. Return periods of 1–25 years in RCP8.5 also show moderate drought severity across most of the watershed area.

For SDI-6, we found that return periods of 10–50 years exhibit the greatest drought severity, with absolute SDI values ranging from 2.24 to 2.74. These severe droughts cover longer periods and have a relatively long duration. In shorter return periods of 1–10 years, fewer droughts are observed. In the RCP4.5 scenario, severe droughts are primarily seen in 25–50 year return periods, with absolute SDI values ranging from 2.31 to 3.76, indicating continuous drought severity over a long period. For return periods of 0–10 years, no droughts occur in the first 5 years, and moderate droughts are observed in the 5–10 year period, with absolute SDI values ranging from 1.56 to 1.97. However, the RCP8.5 scenario shows a more extensive and severe drought distribution. In return periods of 10–50 years, high severity covers a wide area of the watershed, with absolute SDI values ranging from 2.38 to 4.38. For return periods of 1–10 years, a moderate level of drought severity covers most of the watershed area.

We found that the probability of severe hydrological drought in the LCKW is quite high due to low annual recurrence. The severity of drought varies across different periods, showing distinct behavior at different recurrence intervals. In all scenarios for SDI-3 and SDI-6, drought severity generally increases with longer return periods, except for SDI-3 in the RCP4.5 scenario, where the most severe droughts are associated with shorter return periods. Notably, SDI-6 exhibits greater severity and duration of droughts than SDI-3, as it better captures the lag between reduced rainfall and its impact on streamflow, providing a clearer indication of hydrological drought conditions.

Overall, the results of this study highlight the significant impact of higher greenhouse gas emissions on drought severity and distribution. The RCP8.5 scenario consistently indicates more severe and widespread drought conditions than the RCP4.5 scenario. The analysis underscores the importance of considering different time scales and return periods when assessing drought severity to understand the potential future impacts in varying climate change scenarios. Effective water resource management and climate mitigation efforts are crucial to addressing the increasing severity and frequency of droughts associated with higher greenhouse gas emissions.

## 4. Discussion

### 4.1. Trends in Future Rainfall

In this study, we analyzed future rainfall patterns in northeastern Thailand in two climate change scenarios, RCP4.5 and RCP8.5, by using simulations based on the EC-Earth3 model. These simulations provide valuable insights into the region's future climate, as confirmed by Pimonsree et al. (2023) [51], who proved their accuracy based on a high spatial correlation coefficient and observational rainfall data across Southeast Asia. The findings suggest a significant increase in both annual and seasonal rainfall, particularly during the rainy season from May to November [52,53]. Despite the overall increase in rainfall, the pattern of daily rainfall remains stable, which is consistent with the findings of other studies [54].

First, in the RCP4.5 scenario, researchers such as Tammadid et al. (2023) [55] and Boonwicahi et al. (2018) [56] predicted a significant increase in rainfall in northeastern Thailand by the period 2030–2035, with annual precipitation expected to increase by 13%. Additionally, for the nearby watershed area, Li et al. (2021) [57] indicated that rainfall during the wet season is projected to increase substantially in RCP4.5, which is particularly crucial for agriculture and water resource management in the region. These changes are attributed to the warming of global climate, as noted by the IPCC, which forecasted an intensified water cycle in Southeast Asia [58]. This scenario also suggests higher surface temperatures and stronger winds, potentially leading to more frequent tropical cyclones, increased rainfall, and heightened flood risk [59].

Conversely, in the RCP8.5 scenario, which assumes a higher trajectory of greenhouse gas emissions, the outlook changes significantly. Shrestha et al. (2021) [59] and Okwala et al. (2020) [60] predicted a reduction in rainfall, with a forecasted 11% decrease by 2050 in two watershed areas close to the LCKW, which have the same climate. These nearby watersheds are relevant to our study because they share similar hydrological and meteorological characteristics, making them a useful proxy for understanding potential impacts in our primary study area. The reduction in rainfall is more pronounced during the wet seasons, which could critically impact the region's hydrology. The decrease in rainfall, combined with expanded irrigation practices and the increasing frequency and intensity of El Niño events, suggests that streamflow and water availability will be significantly affected, potentially leading to more severe drought conditions and water scarcity [61]. The results of the examination of future rainfall trends in these two distinct climate scenarios, RCP4.5 and RCP8.5, highlight significant variations in regional climate responses. This underscores the importance of considering the impacts of climate change on water management in northeastern Thailand.

### 4.2. *Effects on Future Streamflow*

This study assessed the impacts of climate change on streamflow, focusing on how decreases in precipitation during the rainy season contribute significantly to reductions in annual streamflow. In Southeast Asia, our findings are consistent with previous studies, such as Promping et al. (2020) [62], who projected a 3.39–6.15% decrease in streamflow from 2020 to 2050 in the RCP8.5 scenario, with rainy season flows potentially being reduced by 31–47% in the Pasak River Basin, a neighboring watershed. The Pasak River Basin shares similar climatic and hydrological characteristics with our study area, making it a valuable reference point for understanding the broader regional impacts of climate change on water resources.

Conversely, in the RCP4.5 scenario, increased rainfall could lead to significant increases in streamflow. For example, Kimmany et al. (2020) [63] reported 8% and 22% increases in dry season and annual streamflow, respectively. Similarly, Li et al. (2021) [57] projected that streamflow in the Mun River, which is the main river basin of the LCKW, could increase by 10.5%, 20.1%, and 23.2% during 2020–2093, due to high concentrations of greenhouse gases altering, cloud formation, increasing temperatures, and changing precipitation patterns [64]. Increased aerosol levels can lead to more reflective clouds [65], which are less effective in producing rain [66], while weakened atmospheric circulation further reduces storm occurrence, exacerbating water scarcity in the region [67].

However, these increases are not consistent across all regions and are influenced by various factors, including land-use changes, soil moisture retention, and evapotranspiration rates, all of which are impacted by climate change. This complexity suggests that while some regions may experience higher streamflow, others may not see corresponding increases, even with greater rainfall, underscoring the nuanced effects of climate change on hydrology [68]. Given this variability, the accuracy of hydrological model predictions becomes crucial. However, in regions with limited streamflow measurement stations, the scarcity of data can significantly hinder accurate predictions. Therefore, expanding the network of streamflow stations is essential, as it would provide more comprehensive data, thereby improving the accuracy of hydrological models and enhancing climate change projections.

### 4.3. *Characteristics of Hydrological Drought*

The findings suggest that climate change scenarios in RCP8.5 are likely to increase drought severity compared with historical data. Climate change is expected to alter precipitation patterns, leading to more intense and frequent drought events. These results are consistent with previous research, including the study by Satoh et al. (2022) [69], where increased drought severity due to these changes was predicted. Altered precipitation patterns are expected to exacerbate water scarcity in vulnerable areas, as highlighted by Ullah et al. (2023) [70]. While the RCP8.5 scenario projects increased drought severity, the higher emissions associated with this scenario present a greater risk, leading to more widespread and intense droughts. These severe drought conditions could result in substantial ecological impacts, including biodiversity loss and the degradation of water resources [71].

The findings also indicate a high probability of very severe droughts in the Lam Chiang Kri Watershed (LCKW), primarily due to low annual recurrence rates. This finding is supported by Maithong et al. (2022) [45,72], who investigated the spatial distribution of drought return periods in the Mun Watershed, of which the LCKW is a branch watershed. In their study, they found increased drought severity in rivers with high streamflow over extended periods, influenced by natural variations and human activities such as dam operations and water diversions [73,74]. The diverse conditions of the selected rivers provide a comprehensive understanding of drought occurrences across different return periods. Furthermore, the choice of the SDI (e.g., SDI-3 and SDI-6) influences the results of drought behavior. SDI-3, as noted by Hasan et al. (2022) [75], is sensitive to short-term fluctuations and detects more frequent but less severe droughts [76]. In contrast, SDI-6 captures longer-term trends, reflecting more prolonged droughts [77].

While natural factors drive drought occurrences, human activities, such as land-use change and water extraction, can further exacerbate their impacts. Therefore, understanding these complexities is crucial to developing effective water management and adaptation strategies to mitigate the socioeconomic consequences of drought [78]. Although focusing on the specific years 2029 and 2039 provides valuable insights into near-term and mid-term climate impacts, this approach has its limitations. Specifically, single-year projections can effectively highlight extremes but are also susceptible to anomalies and may not fully capture long-term trends. Consequently, to enhance the reliability of future research, it would be beneficial to analyze broader periods spanning multiple decades. Such an approach would offer a more stable assessment of drought trends, thereby providing a comprehensive understanding that better informs water resource management and adaptation strategies.

### *4.4. Management Implications and Future Perspectives*

Based on the results of this study, we suggest using Reservoir Operation Study (ROS) technology to optimize water storage and drainage management, particularly in the northeastern region, where unique geographical challenges complicate water management [79]. ROS technology is particularly effective in addressing these issues by enabling more precise control of water resources, ensuring that water is available when needed for agriculture, and reducing the risk of both water shortages and flooding. Additionally, farmers are encouraged to shift to less water-intensive crops and zone their cultivation based on soil and water availability. Sustainable farming practices, such as alternating wet and dry farming, using fertilizers, avoiding burning stubble and rice straw, and integrating pest management, are crucial to reducing greenhouse gas emissions and improving rice farming efficiency [80].

In the context of drought preparedness, particularly in regions such as Thailand, adopting integrated water resource management (IWRM) principles can enhance water management efficiency [81]. Developing and planting drought-resistant crops, which require less water and are more resilient to dry conditions, can significantly reduce agriculture's vulnerability to drought [82]. Empowering local communities in drought management is also essential, as community-based approaches integrate local knowledge into preparedness and response strategies.

The methodology proposed in this study can be applied globally, especially in regions such as the one studied, facilitating strategic drought management. This involves engaging stakeholders, policymakers, and water resource managers in monitoring, prediction, modeling, and disaster risk reduction. However, we acknowledge the uncertainties of this study, such as limited observational streamflow data, which affected model accuracy. Future research should incorporate changes in climate variables, land use, soil conditions, and population growth, especially in agricultural regions reliant on irrigation.

By investing in research and data collection, we can improve our ability to predict future changes and inform decision-making processes, ensuring more resilient and sustainable water and agricultural management practices. Finally, continued research and data collection are essential to refine our understanding of climate change impacts and develop effective adaptation and mitigation strategies.

## 5. Conclusions

Climate change impacts in Thailand, particularly in the LCKW region, are expected to vary significantly. In RCP4.5, an increase in annual and seasonal rainfall, especially during the rainy season, is projected, while RCP8.5 predicts a decrease, particularly in wet seasons, potentially leading to water scarcity. Streamflow projections show potential increases in RCP4.5, but a significant decrease is anticipated in RCP8.5.

The findings suggest that future droughts in RCP8.5 scenarios may be more intense and frequent compared with historical periods. Severe droughts are likely to occur more frequently and with greater intensity in RCP8.5. The SDI-3 analysis indicates quick-developing, short-duration droughts, whereas the SDI-6 analysis shows more widespread and prolonged drought conditions, especially in RCP8.5.

These projections underscore the urgent need for the implementation of proactive water management strategies in the LCKW, such as expanding reservoir capacity, improving irrigation efficiency, and promoting water conservation. The expected changes in rainfall and drought patterns will likely have significant socio-economic impacts, particularly on agriculture and water supply. Policymakers should prioritize investments in early warning systems, drought-resistant crops, and community-based adaptation to enhance resilience and ensure water security.

**Author Contributions:** Conceptualization, methodology, analysis, investigation, visualization, and writing—original draft, S.C.; supervision, review, and editing, P.T. and N.K. All authors have read and agreed to the published version of the manuscript.

**Funding:** This research is funded by Kasetsart University through the Graduate School Fellowship Program.

**Institutional Review Board Statement:** Not applicable.

**Informed Consent Statement:** Not applicable.

**Data Availability Statement:** The datasets, including technical and observational data, generated and/or analyzed during this study are available from the corresponding author upon reasonable request.

**Acknowledgments:** The authors are grateful to the Graduate School, Kasetsart University, for their support through the Graduate School Fellowship Program. We also thank the Thai Meteorological Department for providing climate data, including the downscaled CMIP5 GCM, and the Royal Irrigation Department for supplying hydrological data.

**Conflicts of Interest:** The authors declare no conflicts of interest.

## References

1. Swain, D.L.; Singh, D.; Touma, D.; Diffenbaugh, N.S. Attributing Extreme Events to Climate Change: A New Frontier in a Warming World. *One Earth* **2020**, *2*, 522–527. [CrossRef]
2. Bates, B.; Kundzewicz, Z.W.; Wu, S.; Burkett, V.; Doell, P.; Gwary, D.; Hanson, C.; Heij, B.; Jiménez, B.; Kaser, G.; et al. Climate Change and Water. *Tech. Pap. Intergov. Panel Clim. Chang.* **2008**. [CrossRef]
3. Wilhite, D.A.; Glantz, M.H. Understanding: The Drought Phenomenon: The Role of Definitions. *Water Int.* **1985**, *10*, 111–120. [CrossRef]
4. Secci, D.; Tanda, M.G.; D'Oria, M.; Todaro, V.; Fagandini, C. Impacts of Climate Change on Groundwater Droughts by Means of Standardized Indices and Regional Climate Models. *J. Hydrol.* **2021**, *603*, 127154. [CrossRef]
5. Jehanzaib, M.; Shah, S.A.; Yoo, J.-Y.; Kim, T.-W. Investigating the Impacts of Climate Change and Human Activities on Hydrological Drought Using Non-Stationary Approaches. *J. Hydrol.* **2020**, *588*, 125052. [CrossRef]
6. Vargas, J.; Paneque, P. Challenges for the Integration of Water Resource and Drought-Risk Management in Spain. *Sustainability* **2019**, *11*, 3356. [CrossRef]
7. Jahangir, M.H.; Yarahmadi, Y. Hydrological Drought Analyzing and Monitoring by Using Streamflow Drought Index (SDI) (Case Study: Lorestan, Iran). *Arab. J. Geosci.* **2020**, *13*, 110. [CrossRef]

8. Meilutytė-Lukauskienė, D.; Nazarenko, S.; Kobets, Y.; Akstinas, V.; Sharifi, A.; Haghighi, A.T.; Hashemi, H.; Kokorīte, I.; Ozolina, B. Hydro-Meteorological Droughts Across the Baltic Region: The Role of the Accumulation Periods. *Sci. Total Environ.* **2024**, *913*, 169669. [CrossRef]
9. Athukoralalage, D.; Brookes, J.; McDowell, R.W.; Mosley, L.M. Impact of Hydrological Drought Occurrence, Duration, and Severity on Murray-Darling Basin Water Quality. *Water Res.* **2024**, *252*, 121201. [CrossRef]
10. Wubneh, M.A.; Alemu, M.G.; Fekadie, F.T.; Worku, T.A.; Demamu, M.T.; Aman, T.F. Meteorological and Hydrological Drought Monitoring and Trend Analysis for Selected Gauged Watersheds in the Lake Tana Basin, Ethiopia: Under Future Climate Change Impact Scenario. *Sci. Afr.* **2023**, *20*, e01738. [CrossRef]
11. Malik, A.; Kumar, A.; Singh, R.P. Application of Heuristic Approaches for Prediction of Hydrological Drought Using Multi-Scalar Streamflow Drought Index. *Water Resour. Manag.* **2019**, *33*, 3985–4006. [CrossRef]
12. Katipoğlu, O.M. Prediction of Streamflow Drought Index for Short-Term Hydrological Drought in the Semi-Arid Yesilirmak Basin Using Wavelet Transform and Artificial Intelligence Techniques. *Sustainability* **2023**, *15*, 4507. [CrossRef]
13. Boonwichai, S.; Shrestha, S.; Pradhan, P.; Babel, M.S.; Datta, A. Adaptation Strategies for Rainfed Rice Water Management under Climate Change in Songkhram River Basin, Thailand. *J. Water Clim. Chang.* **2021**, *12*, 2181–2198. [CrossRef]
14. Sangkhaphan, S.; Shu, Y. The Effect of Rainfall on Economic Growth in Thailand: A Blessing for Poor Provinces. *Economies* **2020**, *8*, 37. [CrossRef]
15. Intergovernmental Panel on Climate Change (IPCC). Observations: Atmosphere and Surface. In *Climate Change 2013—The Physical Science Basis: Working Group I Contribution to the Fifth Assessment Report of the Intergovernmental Panel on Climate Change*; Cambridge University Press: Cambridge, UK, 2014; pp. 159–254.
16. Hormwichian, R.; Kaewplang, S.; Kangrang, A.; Supakosol, J.; Boonrawd, K.; Sriworamat, K.; Muangthong, S.; Songsaengrit, S.; Prasanchum, H. Understanding the Interactions of Climate and Land Use Changes with Runoff Components in Spatial-Temporal Dimensions in the Upper Chi Basin, Thailand. *Water* **2023**, *15*, 2022. [CrossRef]
17. Liu, S.; Chen, J.; Li, J.; Li, T.; Shi, H.; Sivakumar, B. Building a Large Dam: Identifying the Relationship Between Catchment Area and Slope Using the Confidence Ellipse Approach. *Geosci. Lett.* **2023**, *10*, 4. [CrossRef]
18. Dlamini, T.; Songsom, V.; Koedsin, W.; Ritchie, R.J. Intensity, Duration, and Spatial Coverage of Aridity During Meteorological Drought Years Over Northeast Thailand. *Climate* **2022**, *10*, 35. [CrossRef]
19. Prasanchum, H.; Pimput, Y. Risk Assessment of Flash Flood Situation Under Land Use Change Using Daily SWAT Streamflow Simulation in Loei Basin, Northeastern Thailand. *IOP Conf. Ser. Earth Environ. Sci.* **2023**, *1151*, 012015. [CrossRef]
20. Plaiklang, S.; Sutthivanich, I.; Sritarapipat, T.; Panurak, K.; Ogawa, S.; Charungthanakij, S.; Maneewan, U.; Thongrueang, N. Desertification Assessment Using MEDALUS Model in Upper Lamchiengkrai Watershed, Thailand. *Int. Arch. Photogramm. Remote Sens. Spatial Inf. Sci.* **2020**, *XLIII*, 1257–1262. [CrossRef]
21. Zhang, T.; Su, X.; Zhang, G.; Wu, H.; Liu, Y. Projections of the Characteristics and Probability of Spatially Concurrent Hydrological Drought in a Cascade Reservoirs Area Under CMIP6. *J. Hydrol.* **2022**, *613*, 128472. [CrossRef]
22. Royal Irrigation Department. Developing Water Resources to Solve Drought and Flood Problems in the Lam Chiang Krai Watershed. Available online: https://www.undp.org/thailand/press-releases/undp-and-royal-irrigation-department-thailand-marks-world-water-day-2023-new-gcf-funded-project (accessed on 23 April 2023).
23. Nabinejad, S.; Schüttrumpf, H. Flood Risk Management in Arid and Semi-Arid Areas: A Comprehensive Review of Challenges, Needs, and Opportunities. *Water* **2023**, *15*, 1070. [CrossRef]
24. Northeastern Meteorological Center (Lower Part). Daily Meteorological Data, Including Precipitation, Temperature, Humidity, Wind Speed, and Radiation. Available online: https://ubonmet.tmd.go.th/ (accessed on 20 April 2022).
25. Yamada, I. Thiessen Polygons. In *International Encyclopedia of Geography*; Wiley: Hoboken, NJ, USA, 2016; pp. 1–6.
26. Royal Irrigation Department. Thailand Daily Hydrological Data, Including Streamflow. Available online: https://water.rid.go.th/hydrology/ (accessed on 20 April 2023).
27. National Aeronautics and Space Administration (NASA). Earth Science Data Systems (ESDS) Program. Available online: https://search.asf.alaska.edu/#/ (accessed on 10 January 2022).
28. Kamworapan, S.; Surussavadee, C. Evaluation of CMIP5 Global Climate Models for Simulating Climatological Temperature and Precipitation for Southeast Asia. *Adv. Meteorol.* **2019**, *2019*, 1067365. [CrossRef]
29. Chung, J.X.; Juneng, L.; Santisirisomboon, G.; Ngo-Duc, T.; Phan-Van, T.; Trinh-Tuan, L.; Cruz, F.; Dado, J.; Srisawadwong, R.; Gunawan, D.; et al. Future Changes in Mean and Extreme Precipitation Over Peninsular Malaysia Using CORDEX-SEA 5 km Simulations. *APN Sci. Bull.* **2023**, *13*, 263–276. [CrossRef]
30. Lee, Z.H.; Sethupathi, S.; Lee, K.T.; Bhatia, S.; Mohamed, A.R. An Overview on Global Warming in Southeast Asia: $CO_2$ Emission Status, Efforts Done, and Barriers. *Renew. Sustain. Energy Rev.* **2013**, *28*, 71–81. [CrossRef]
31. Sentian, J.; Payus, C.M.; Herman, F.; Kong, V.W.Y. Climate Change Scenarios Over Southeast Asia. *APN Sci. Bull.* **2022**, *12*, 102–122. [CrossRef]
32. Schwalm, C.R.; Glendon, S.; Duffy, P.B. RCP8.5 Tracks Cumulative $CO_2$ Emissions. *Proc. Natl. Acad. Sci. USA* **2020**, *117*, 19656–19657. [CrossRef]
33. Thomson, A.M.; Calvin, K.V.; Smith, S.J.; Kyle, G.P.; Volke, A.; Patel, P.; Delgado-Arias, S.; Bond-Lamberty, B.; Wise, M.A.; Clarke, L.E.; et al. RCP4.5: A Pathway for Stabilization of Radiative Forcing by 2100. *Clim. Chang.* **2011**, *109*, 77–94. [CrossRef]

34. Kotamarthi, R.; Hayhoe, K.; Mearns, L.O.; Wuebbles, D.; Jacobs, J.; Jurado, J. Dynamical Downscaling. In *Downscaling Techniques for High-Resolution Climate Projections: From Global Change to Local Impacts*; Cambridge University Press: Cambridge, UK, 2021; pp. 64–81.
35. Zhou, Q.; Chen, D.; Hu, Z.-Y.; Chen, X. Decompositions of Taylor Diagram and DISO Performance Criteria. *Int. J. Climatol.* **2021**, *41*, 4195–4210. [CrossRef]
36. Gassman, P.W.; Sadeghi, A.M.; Srinivasan, R. Applications of the SWAT Model Special Section: Overview and Insights. *J. Environ. Qual.* **2014**, *43*, 1–8. [CrossRef]
37. Abbaspour, K.C.; Yang, J.; Maximov, I.; Siber, R.; Bogner, K.; Mieleitner, J.; Zobrist, J.; Srinivasan, R. Modelling Hydrology and Water Quality in the Pre-Alpine/Alpine Thur Watershed Using SWAT. *J. Hydrol.* **2007**, *333*, 413–430. [CrossRef]
38. Aawar, T.; Khare, D. Assessment of Climate Change Impacts on Streamflow through Hydrological Model Using SWAT Model: A Case Study of Afghanistan. *Model. Earth Syst. Environ.* **2020**, *6*, 1427–1437. [CrossRef]
39. Moriasi, D.N.; Arnold, J.G.; Van Liew, M.W.; Bingner, R.L.; Harmel, R.D.; Veith, T.L. Model Evaluation Guidelines for Systematic Quantification of Accuracy in Watershed Simulations. *Trans. ASABE* **2007**, *50*, 885–900. [CrossRef]
40. Nash, J.E.; Sutcliffe, J.V. River Flow Forecasting Through Conceptual Models Part I—A Discussion of Principles. *J. Hydrol.* **1970**, *10*, 282–290. [CrossRef]
41. Arnold, J.G.; Moriasi, D.N.; Gassman, P.W.; Abbaspour, K.C.; White, M.J.; Srinivasan, R.; Santhi, C.; Harmel, R.D.; van Griensven, A.; Van Liew, M.W.; et al. SWAT: Model Use, Calibration, and Validation. *Trans. ASABE* **2012**, *55*, 1491–1508. [CrossRef]
42. Gupta, H.V.; Sorooshian, S.; Yapo, P.O. Status of Automatic Calibration for Hydrologic Models: Comparison with Multilevel Expert Calibration. *J. Hydrol. Eng.* **1999**, *4*, 135–143. [CrossRef]
43. Gupta, H.V.; Kling, H.; Yilmaz, K.K.; Martinez, G.F. Decomposition of the Mean Squared Error and NSE Performance Criteria: Implications for Improving Hydrological Modelling. *J. Hydrol.* **2009**, *377*, 80–91. [CrossRef]
44. Nalbantis, I.; Tsakiris, G. Assessment of Hydrological Drought Revisited. *Water Resour. Manag.* **2009**, *23*, 881–897. [CrossRef]
45. McKee, T.B.; Doesken, N.J.; Kleist, J. The Relationship of Drought Frequency and Duration to Time Scales. In *Preprints, 8th Conference on Applied Climatology*; American Meteorological Society: Boston, MA, USA, 1993; pp. 179–184.
46. Yevjevich, V.M. *An Objective Approach to Definitions and Investigations of Continental Hydrologic Droughts*; Colorado State University: Fort Collins, CO, USA, 1967; Volume 23.
47. Guo, H.; Bao, A.; Liu, T.; Ndayisaba, F.; He, D.; Kurban, A.; De Maeyer, P. Meteorological Drought Analysis in the Lower Mekong Basin Using Satellite-Based Long-Term CHIRPS Product. *Sustainability* **2017**, *9*, 901. [CrossRef]
48. Ghabelnezam, E.; Mostafazadeh, R.; Hazbavi, Z.; Huang, G. Hydrological Drought Severity in Different Return Periods in Rivers of Ardabil Province, Iran. *Sustainability* **2023**, *15*, 2547. [CrossRef]
49. Altunkaynak, A.; Wang, K.-H. Triple Diagram Models for Prediction of Suspended Solid Concentration in Lake Okeechobee, Florida. *J. Hydrol.* **2010**, *387*, 165–175. [CrossRef]
50. Mostafazadeh, R.; Shahabi, M.; Zabihi, M. Analysis of Meteorological Drought Using Triple Diagram Model in the Kurdistan Province, Iran. *Geogr. Plan. Space Q. J.* **2015**, *5*, 129–140.
51. Pimonsree, S.; Kamworapan, S.; Gheewala, S.H.; Thongbhakdi, A.; Prueksakorn, K. Evaluation of CMIP6 GCMs Performance to Simulate Precipitation Over Southeast Asia. *Atmos. Res.* **2023**, *282*, 106522. [CrossRef]
52. Tangang, F.; Santisirisomboon, J.; Juneng, L.; Salimun, E.; Chung, J.; Supari, S.; Cruz, F.; Ngai, S.T.; Ngo-Duc, T.; Singhruck, P.; et al. Projected Future Changes in Mean Precipitation Over Thailand Based on Multi-Model Regional Climate Simulations of CORDEX Southeast Asia. *Int. J. Climatol.* **2019**, *39*, 5413–5436. [CrossRef]
53. Maimone, M.; Malter, S.; Anbessie, T.; Rockwell, J. Three Methods of Characterizing Climate-Induced Changes in Extreme Rainfall: A Comparison Study. *J. Water Clim. Chang.* **2023**, *14*, 4245–4260. [CrossRef]
54. Khalil, A. Combined Use of Graphical and Statistical Approaches for Rainfall Trend Analysis in the Mae Klong River Basin, Thailand. *J. Water Clim. Chang.* **2023**, *14*, 4642–4668. [CrossRef]
55. Tammadid, W.; Nantasom, K.; Sirksiri, W.; Vanitchung, S.; Promjittiphong, C.; Limsakul, A.; Hanpattanakit, P. Future Projections of Precipitation and Temperature in Northeast Thailand Using Bias-Corrected Global Climate Models. *Chiang Mai J. Sci.* **2023**, *50*, 1–8. [CrossRef]
56. Boonwichai, S.; Shrestha, S. Climate Change Impact on Rained Rice Production and Irrigation Water Requirement in Songkhram River Basin, Thailand. *Int. J. Eng. Technol. Sci.* **2018**, *5*, 94–104.
57. Li, C.; Fang, H. Assessment of Climate Change Impacts on the Streamflow for the Mun River in the Mekong Basin, Southeast Asia: Using SWAT Model. *CATENA* **2021**, *201*, 105199. [CrossRef]
58. Allen, S.K.; Barros, V.; Burton, I.; Campbell-Lendrum, D.; Cardona, O.-D.; Cutter, S.L.; Dube, O.P.; Ebi, K.L.; Field, C.B.; Handmer, J.W.; et al. Summary for Policymakers. In *Managing the Risks of Extreme Events and Disasters to Advance Climate Change Adaptation: Special Report of the Intergovernmental Panel on Climate Change*; Cambridge University Press: Cambridge, UK, 2012; pp. 3–22.
59. Shrestha, S.; Roachanakanan, R. Extreme Climate Projections Under Representative Concentration Pathways in the Lower Songkhram River Basin, Thailand. *Heliyon* **2021**, *7*, e06146. [CrossRef]
60. Okwala, T.; Shrestha, S.; Ghimire, S.; Mohanasundaram, S.; Datta, A. Assessment of Climate Change Impacts on Water Balance and Hydrological Extremes in Bang Pakong-Prachin Buri River Basin, Thailand. *Environ. Res.* **2020**, *186*, 109544. [CrossRef]
61. Orkodjo, T.P.; Kranjac-Berisavijevic, G.; Abagale, F.K. Impact of Climate Change on Future Precipitation Amounts, Seasonal Distribution, and Streamflow in the Omo-Gibe Basin, Ethiopia. *Heliyon* **2022**, *8*, e09711. [CrossRef] [PubMed]

62. Promping, T.; Tingsanchali, T.; Chuanpongpanich, S. Impacts of Future Climate Change on Inflow to Pasak Jolasid Dam in Pasak River Basin, Thailand. *Water Environ. J.* **2021**, *35*, 1355–1369.
63. Kimmany, B.; Visessri, S.; Pech, P.; Ekkawatpanit, C. The Impact of Climate Change on Hydro-Meteorological Droughts in the Chao Phraya River Basin, Thailand. *Water* **2024**, *16*, 100. [CrossRef]
64. Tuccella, P.; Menut, L.; Briant, R.; Deroubaix, A.; Khvorostyanov, D.; Mailler, S.; Siour, G.; Turquety, S. Implementation of Aerosol-Cloud Interaction Within WRF-CHIMERE Online Coupled Model: Evaluation and Investigation of the Indirect Radiative Effect from Anthropogenic Emission Reduction on the Benelux Union. *Atmosphere* **2019**, *10*, 543. [CrossRef]
65. Gryspeerdt, E.; Smith, T.W.P.; O'Keeffe, E.; Christensen, M.W.; Goldsworth, F.W. The Impact of Ship Emission Controls Recorded by Cloud Properties. *Geophys. Res. Lett.* **2019**, *46*, 12547–12555. [CrossRef]
66. Rosenfeld, D.; Lohmann, U.; Raga, G.; O'Dowd, C.; Kulmala, M.; Sandro, F.; Reissell, A.; Andreae, M. Flood or Drought: How Do Aerosols Affect Precipitation? *Science* **2008**, *321*, 1309–1313. [CrossRef]
67. Intergovernmental Panel on Climate Change (IPCC). *Climate Change 2021—The Physical Science Basis: Working Group I Contribution to the Sixth Assessment Report of the Intergovernmental Panel on Climate Change*; Cambridge University Press: Cambridge, UK, 2023.
68. Kabir, M.; Habiba, U.E.; Khan, W.; Shah, A.; Rahim, S.; Rios-Escalante, P.R.D.L.; Farooqi, Z.-U.-R.; Ali, L.; Shafiq, M. Climate Change Due to Increasing Concentration of Carbon Dioxide and Its Impacts on Environment in 21st Century: A Mini Review. *J. King Saud. Univ. Sci.* **2023**, *35*, 102693. [CrossRef]
69. Satoh, Y.; Yoshimura, K.; Pokhrel, Y.; Kim, H.; Shiogama, H.; Yokohata, T.; Hanasaki, N.; Wada, Y.; Burek, P.; Byers, E.; et al. The Timing of Unprecedented Hydrological Drought Under Climate Change. *Nat. Commun.* **2022**, *13*, 3287. [CrossRef]
70. Ullah, I.; Xinmin, Z.; Hina, S.; Syed, S.; Ma, X.; Iyakaremye, V.; Yin, J.; Singh, V. Recent and Projected Changes in Water Scarcity and Unprecedented Drought Events Over Southern Pakistan. *Front. Earth Sci.* **2023**, *11*, 977460.
71. Xu, Y.; Zhang, X.; Hao, Z.; Hao, F.; Li, C. Systematic Assessment of the Development and Recovery Characteristics of Hydrological Drought in a Semi-Arid Area. *Sci. Total Environ.* **2022**, *836*, 155472. [CrossRef]
72. Maithong, P.; Mapiam, P.P.; Lipiwattanakarn, S. Spatial Distribution of Drought Return Periods for the Mun Basin Using a Bivariate Copula Method. *J. King Mongkut's Univ. Technol. North Bangk.* **2018**, *28*, 126–134. [CrossRef]
73. Akbari, M.; Mirchi, A.; Roozbahani, A.; Gafurov, A.; Kløve, B.; Haghighi, A.T. Desiccation of the Transboundary Hamun Lakes Between Iran and Afghanistan in Response to Hydro-Climatic Droughts and Anthropogenic Activities. *J. Great Lakes Res.* **2022**, *48*, 876–889. [CrossRef]
74. Parchami, N.; Mostafazadeh, R.; Ouri, A.E.; Imani, R. Spatio-Temporal Analysis of River Flow Master Recession Curve Characteristics Over Ardabil Province Rivers. *Phys. Chem. Earth Parts A/B/C* **2024**, *134*, 103586. [CrossRef]
75. Hasan, H.H.; Razali, S.F.; Muhammad, N.S.; Ahmad, A. Modified Hydrological Drought Risk Assessment Based on Spatial and Temporal Approaches. *Sustainability* **2022**, *14*, 11235. [CrossRef]
76. Fung, K.F.; Huang, Y.F.; Koo, C.H. Investigation of Streamflow as a Seasonal Hydrological Drought Indicator for a Tropical Region. *Water Supply* **2020**, *20*, 609–620. [CrossRef]
77. Tareke, K.A.; Awoke, A.G. Comparing Surface Water Supply Index and Streamflow Drought Index for Hydrological Drought Analysis in Ethiopia. *Heliyon* **2022**, *8*, e12000. [CrossRef]
78. Fabian, P.S.; Kwon, H.-H.; Vithanage, M.; Lee, J.-H. Modeling, Challenges, and Strategies for Understanding Impacts of Climate Extremes (Droughts and Floods) on Water Quality in Asia: A Review. *Environ. Res.* **2023**, *225*, 115617. [CrossRef]
79. Wu, X.; Shen, X.; Wei, C.; Xie, X.; Li, J. Reservoir Operation Sequence- and Equity Principle-Based Multi-Objective Ecological Operation of Reservoir Group: A Case Study in a Basin of Northeast China. *Sustainability* **2022**, *14*, 10826. [CrossRef]
80. Shang, Z.; Abdalla, M.; Xia, L.; Zhou, F.; Sun, W.; Smith, P. Can Cropland Management Practices Lower Net Greenhouse Emissions Without Compromising Yield? *Glob. Chang. Biol.* **2021**, *27*, 4657–4670. [CrossRef]
81. Prakongsri, P.; Santiboon, T. Effective Water Resources Management for Communities in the Chi River Basin in Thailand. *Environ. Claims J.* **2020**, *32*, 323–348. [CrossRef]
82. Begna, T. Impact of Drought Stress on Crop Production and Its Management Options. *Int. J. Res. Agron.* **2021**, *4*, 66–74. [CrossRef]

*atmosphere*

*Article*

# Critical Drought Characteristics: A New Concept Based on Dynamic Time Period Scenarios

**Ahmad Abu Arra [1,2,*], Mehmet Emin Birpınar [3], Şükrü Ayhan Gazioğlu [1] and Eyüp Şişman [1,4,*]**

1 Department of Civil Engineering, Yildiz Technical University, Istanbul 34220, Türkiye
2 Department of Civil and Architectural Engineering, An-Najah National University, Nablus 44830, Palestine
3 Ministry of Environment and Urbanization, Ankara 06510, Türkiye
4 Department of Civil Engineering, School of Engineering and Natural Sciences, Istanbul Medipol University, Istanbul 34810, Türkiye
* Correspondence: ahmad.arra@std.yildiz.edu.tr (A.A.A.); esisman@yildiz.edu.tr or esisman@medipol.edu.tr (E.Ş.); Tel.: +90-5010318307 (A.A.A.)

**Abstract:** In research on monitoring drought events, analysis is often carried out using a single period as a reference. On the other hand, changing this default period in drought calculations causes the drought index values obtained from research to differ. As a gap in the literature, this point highlights the necessity of investigating the effect of various time periods on drought characteristics. It underscores the need to propose a new concept and methodology to address this gap effectively. This research aims to analyze critical drought characteristics through dynamic time period scenarios. For the first time in the literature, drought indices and potential and critical characteristics were analyzed for various (dynamic) time periods. Drought analysis was carried out for 13 time period scenarios with 10-year intervals from a meteorological station in Durham (1872–2021) by changing the initial time condition using the Standardized Precipitation Index (SPI). The results showed that in addition to the similarities, there are significant differences between drought characteristics. For example, in some time period scenarios, a drought event was recorded during a specific period, while in other scenarios (S5–S7, S10–S13), no drought was detected during the same period, like in SPI 1. Additionally, for SPI 12, the drought duration varied significantly, lasting between 20 and 29 months, and for SPI 6, the drought duration varied between 3 and 13 months. Regarding the intensity, SPI 1 ranged between −0.89 and −1.33, indicating a 33% difference, and the SPI 3 intensity ranged between −1.08 and −1.91, indicating a 50% increase in intensity. This research significantly contributes to the field by providing a novel approach using dynamic time period scenarios to determine critical drought characteristics, offering valuable insights for water resource management, drought mitigation planning, and design purposes.

**Keywords:** potential drought characteristics; critical drought; drought evaluation; standardized precipitation index (SPI); dynamic

## 1. Introduction

Precipitation plays a crucial role in the hydrological cycle, impacting agricultural productivity, regional climate, and broader interactions within atmospheric and terrestrial systems [1,2]. It serves as a key indicator for evaluating the effects of climate change, directly affecting variables like soil moisture levels, streamflow, and groundwater replenishment [3–6]. Global climate changes, which are among the most significant environmental challenges, are resulting in the heightened frequency and severity of extreme weather events, notably droughts [7,8]. Recent research, such as that carried out by Gu et al. [9], anticipates a substantial increase in both the intensity and socioeconomic vulnerability of worldwide droughts under climates warmer by 1.5 and 2 °C, confirming the increasing imperative of understanding and mitigating drought effects. Droughts, observed in diverse climatic conditions, have extensive consequences in many sectors, such as agriculture, energy, water resources, and ecosystems [10,11]. A recent report from the United Nations on World Water Development [12] indicates that approximately four billion people worldwide experience water scarcity for at least one month annually. According to the World Economic Forum [13], economic losses from climate-related disasters, including droughts, amounted to nearly $1.5 trillion in the decade leading up to 2019. Also, Africa's drought-related economic losses in the past 50 years have been about 70 billion USD [14]. Considered among the most catastrophic natural events globally according to their widespread geographical impact [15], understanding the complex temporal patterns of droughts is vital for proficient water resource management and the formulation of strategies to mitigate their effects.

Various standardized drought indices have been employed to evaluate and assess drought. Each index relies on either a singular or multiple hydro-meteorological parameters determining a specific kind of drought, including the Standardized Precipitation Index (SPI) [16], the Standardized Precipitation Evapotranspiration Index (SPEI) [17], the Palmer Drought Severity Index (PDSI) [18], and the Reclamation Drought Index (RDI) [19]. Despite the challenge of determining a universal drought index, the SPI has wide acceptance because of its simplicity and its sole dependence on precipitation data, providing a more straightforward index compared to more complicated indices, and this simplicity makes the SPI particularly well suited to regions with restricted data availability [20]. Its efficiency across diverse time scales has been demonstrated in numerous studies, confirming the SPI's vital role in drought evaluation, particularly under the urgent considerations associated with global climate change [21–23].

After a thorough evaluation of the literature and the calculation and use of drought indices, it is noticed that these indices are generally used and performed for a single time period, determined by taking the available time period of the data as a reference [24–26]. Abu Arra and Şişman [24] analyzed the difference between the used time period and an acceptable time period using different statical metrics and indicated that, for meteorological droughts, a 10-year period and, for hydrological droughts, a 20-year period can be used with high confidence to yield acceptable results for the used time period. It is seen that these time periods vary in research conducted in the same or proximate study areas. For example, Gumus [27] evaluated drought in Türkiye using the SPI method for the period between 1970 and 2021, while Dabanlı et al. [28] evaluated drought in Türkiye using SPI for the period between 1931 and 2010, leading to different results. However, the research topics, study regions, and used methodologies overlap significantly. It is noticeable that this difference in basic assumptions, which is often ignored in analysis, calculations, and evaluations, brings incomplete and erroneous results. Therefore, the index calculations in drought analysis using a single time period selected in the classical way are insufficient for monitoring and evaluating potential and critical drought events that have been and/or may be seen. In order to solve this problem, instead of classical approaches in drought calculations, there is a need to define systematic, innovative concepts and frameworks in which datasets from different periods can be evaluated. Within the framework of the new concept(s) to be developed, potential drought characteristics should be determined according to dynamic time period scenarios, and critical drought characteristics should be determined according to these scenarios.

The aim of this research is to (1) analyze the effect of changing the initial time condition on the drought characteristics, (2) propose a new framework and concepts to identify critical drought characteristics by determining potential drought characteristics based on dynamic time period scenarios instead of the traditional method, and (3) compare the resulting drought characteristics based on different time period scenarios. The calculation of drought characteristics is based on individual drought events. Therefore, the characteristics must be determined for each specific drought event. For this purpose, in this study, drought analysis was carried out with the SPI method for different dynamic time period datasets, prepared by taking precipitation data from the Durham meteorological station (1872–2021) as a reference and application; the drought characteristics were calculated for each period using SPI theory/definitions; and critical drought characteristics were determined for the region in the relevant years.

## 2. Materials and Methods

### *2.1. Application*

The specified methods and determination of the critical drought characteristics using the SPI and dynamic time period scenarios are applied to Durham meteorological station. Durham City is situated in the northeastern part of the United Kingdom. It experiences a hybrid temperate maritime climate characterized by relatively mild summers and cool winters when viewed globally. Typically, July stands out as the warmest month during summer, while January is recognized as the coldest month in winter. The average temperature fluctuates between 5.2 °C (in winter) and 12.5 °C (in summer), with an annual average precipitation of 643 mm (Table 1). The meteorological data for total monthly precipitation (P) span from 1872 to 2021, covering 150 years and originating from the Durham University meteorological station, and have previously been obtained [29]. Notably, the Durham Observatory's weather records are the third longest continuous climate series in the United Kingdom. This dataset was selected because it is a continuous series with a long time period (150 years), it includes high-quality data, and more scenarios can be developed to answer and achieve the purposes of this research in the best way possible. Figure 1 shows the cumulative probability of the original monthly precipitation data and the average monthly time series of the precipitation dataset. Table 1 summarizes the main climatic information of the Durham station, including minimum, maximum, and average monthly precipitation, standard deviation and skewness, and the average and standard deviation of the temperature.

**Table 1.** Climatic information of the Durham station.

| Station's Name | Lat. (N) | Lon. (W) | Average Monthly Precipitation (P)—mm | Standard Deviation (mm) | Min. Monthly Precipitation (mm) | Max. Monthly Precipitation (mm) | Skewness |
|---|---|---|---|---|---|---|---|
| Durham Station | 54.77 | 1.59 | 54.37 | 31.74 | 1.30 | 209.70 | 1.14 |
| | | | Monthly Temperature (T)—°C | | Standard Deviation °C | | |
| | | | 8.6 | | 4.46 | | |

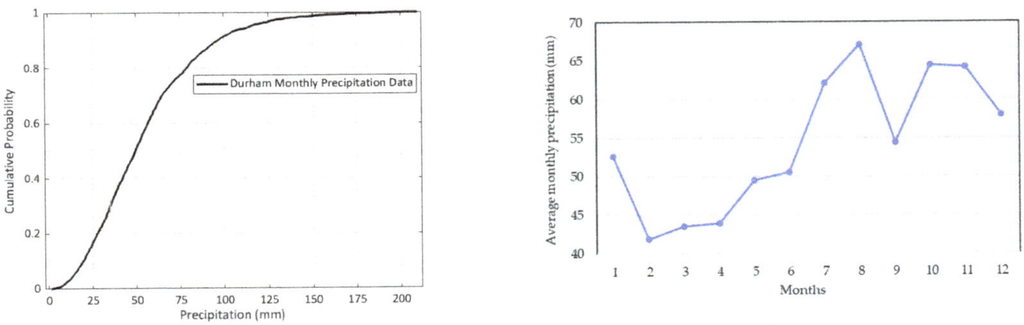


(a) (b)

**Figure 1.** (**a**) Cumulative probability of monthly precipitation data for Durham station (1872–2021), (**b**) average monthly time series of the precipitation dataset between 1872 and 2021.

### *2.2. Homogeneity Tests*

Variations within homogeneous data series are influenced by weather and climate changes. Thus, ensuring the homogeneity of time series data is essential for accurate climate and drought analysis. Using non-homogeneous data in climate research can lead to biased results. Therefore, this study will first test the homogeneity of the data before conducting drought analysis. To assess homogeneity, several absolute homogeneity tests are commonly used, including Pettitt [30], Buishand [31], and the Standard Normal Homogeneity Test (SNHT) proposed by Wijngaard et al. [32]. Homogeneity is evaluated based on the null hypothesis (H0), which assumes no changes within the data. Classifications are then assigned based on the number of tests that accept or reject the null hypothesis as follows:

1. Set 1 is labeled "Homogeneous" if all three methods accept the null hypothesis (H0).
2. Set 2 is labeled "Doubtful" if two of the homogeneity tests accept the null hypothesis (H0).
3. Set 3 is labeled "Suspect" if only one or none of the homogeneity tests accept the null hypothesis (H0).

### 2.3. The Standardized Precipitation Index (SPI)

The SPI calculates the drought index at several time scales, such as 1-month, 3-month, 6-month, and 12-month, depending on the monthly precipitation datasets. The selection of time scales plays a crucial role in determining different types of droughts. Short time scales, such as the commonly used 1-month and 3-month SPIs, indicate meteorological droughts. These shorter durations capture variations in weather patterns and highlight deviations from normal precipitation levels over a relatively brief period. On the other hand, longer time scales, spanning multiple months or even years, provide insights into hydrological droughts [16,20]. As for the method, the original monthly precipitation datasets are fitted to a suitable probability density function. The Gamma PDF has been determined as the best PDF for SPI calculations in most research, as stated by Wang et al. [33]. The selection process for a suitable PDF is carried out using goodness-of-fit tests for the original datasets (precipitation for SPI), including Chi-Square and Kolmogorov–Smirnov [34]. The probabilities are derived through computation of cumulative distribution functions applied to monthly precipitation datasets. Subsequently, these probabilities undergo a probabilistic standardization process, transforming them into a standard index value characterized by a mean of 0 and a standard deviation of 1, as shown in Figure 2. A critical point that must be mentioned is that the probabilistic standardization process differs from statistical standardization. The difference between them in SPI calculation has been studied in detail by Şen and Şişman [25].

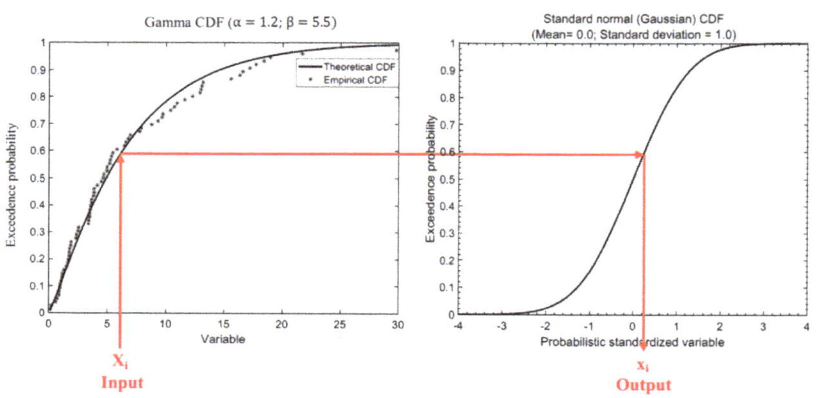


**Figure 2.** Probabilistic standardization process [25].

### 2.4. Drought Characteristics

The initial phase in drought analysis and evaluation involves computing the drought index. Subsequently, three key characteristics, namely duration (D), severity (S), and intensity (I), are derived based on this index. These drought characteristics are calculated using the drought index and based on the drought definition. The run theory introduced by Yavjevich [35] and the SPI theory developed by Mckee et al. [16] are two widely accepted definitions for drought characteristics. Yavjevich [35] states that a drought event commences when the drought index falls below zero and ends when it rises back above zero. Secondly, in identification, Mckee et al. [16] defined drought events (start) using $-1$ instead of 0 thresholds. A notable difference exists in the computation of drought characteristics between run theory and SPI theory. The run theory tends to yield more extreme duration values, while the SPI theory results in more extreme intensity values, as highlighted in [36]. This research uses SPI theory because the higher extreme intensity values are used to calculate the drought characteristics.

The main characteristic and parameter used in this research is drought duration (D), and based on the SPI theory, D is defined as the total number of months when the drought index is less than $-1$ until it returns to a positive value. The second parameter is the summation of the drought index over the drought duration (sum), and the biggest value of the drought indices is called the drought peak. Additionally, dividing (sum) by the drought duration gives the drought intensity. This research also calculates the average and median values for drought intensity for the first time to explore and compare the variations between different time period scenarios. Figure 3 shows the drought index values, drought duration for each drought theory, and drought peak values.

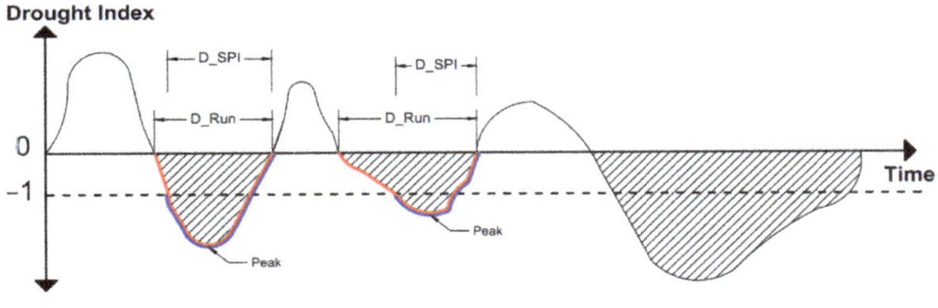


**Figure 3.** Drought duration based on SPI theory (D_SPI, blue line) and run theory (D_Run, red line) and drought index peak value (Peak) relying on drought index values [36].

The main drought characteristics and statistics used in this research are summarized as follows:

1. Drought duration (D) is defined as follows:

$$D_{SPI}\text{ theory} = \text{Number of months between } DI1st_{month} < -1 \text{ and } DI_{\text{until any month returns positive}}. \quad (1)$$

2. Drought severity (sum): Summation of drought index values within the drought duration.
3. Drought intensity (I):

$$\text{Drought intensity} = \text{Drought severity}/\text{Drought duration} \quad (2)$$

4. Median and peak values for the drought event.

The potential drought characteristics are identified as the drought characteristics associated with each selected time period dataset scenario. It contains the drought duration (D), intensity (I), severity (sum), peak, and median values of DI. Considering and using these characteristics, scenarios will be used for specific hydraulic design purposes, such as dam design and water resource and drought management.

### *2.5. Dynamic Time Period Scenarios*

To identify potential and critical drought characteristics, this research employs a methodological approach involving the generation of dynamic time period scenarios, with variations in the initial time conditions. The range of these time periods may include intervals such as 1, 5, or 10 years, changing based on the longest available time period for each condition, the purpose of the drought characteristics, and the climate of the study area. For each time period scenario, the drought index values and associated characteristics are meticulously calculated. Subsequently, a comparative analysis is conducted for specific drought events, considering their characteristics across different time periods. This comparative assessment aims to discern critical and potential drought characteristics by evaluating how the drought index and its related parameters evolve under diverse temporal conditions. Through this systematic exploration of varying time periods and associated drought characteristics, the research aims to contribute valuable insights into understanding the dynamics and critical factors influencing drought events, aiding in more effective drought management and mitigation strategies. Figure 4 shows the methodological approach in this research.

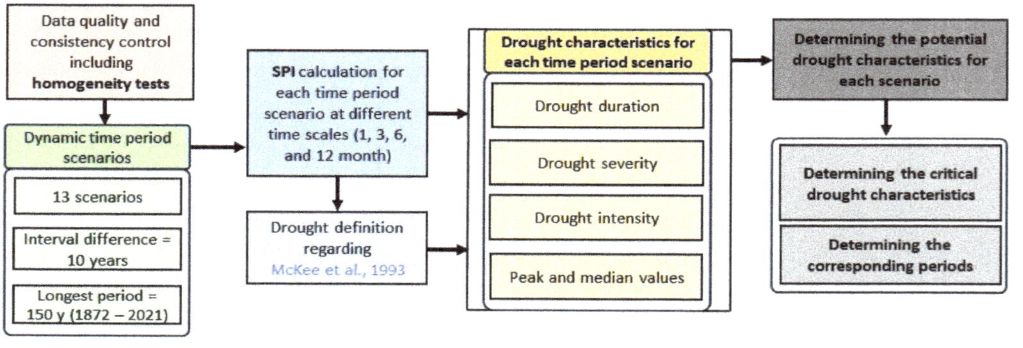


**Figure 4.** Methodological approach [16].

## 3. Results

This research study provides a comprehensive and detailed analysis of Durham station (1872–2021). The SPI drought index values and characteristics are calculated using SPI theory. Based on the longest available period (150 years) and the minimum (ideal) period (30 years), this period is divided into 13 time period scenarios, with 10 years as an interval. The 10-year interval reflects the practical use of the SPI in the literature and leads to 13 time period scenarios which can be analyzed and explained. In the analysis of potential drought characteristics across thirteen different time period scenarios, ranging from 30 to 150 years, and according to the framework of the proposed concept, the results demonstrate significant variability in the response of drought characteristics when assessed over different historical spans. The comparison focuses on the common last 30 years for all scenarios, utilizing the SPI to evaluate the drought duration, intensity, severity, median, and peak values during specific drought events. The time period scenarios with the start and end years for each scenario and the duration for each scenario are presented in Table 2. The results are organized into distinct sub-sections for each time scale to simplify and enhance clarity in the Results section.

**Table 2.** Time period dataset scenarios.

| Scenario | Start and End Year | Time Period Duration (Years) |
|---|---|---|
| S1 | 1992–2021 | 30 |
| S2 | 1982–2021 | 40 |
| S3 | 1972–2021 | 50 |
| S4 | 1962–2021 | 60 |
| S5 | 1952–2021 | 70 |
| S6 | 1942–2021 | 80 |

**Table 2.** *Cont.*

| Scenario | Start and End Year | Time Period Duration (Years) |
|---|---|---|
| S7 | 1932–2021 | 90 |
| S8 | 1922–2021 | 100 |
| S9 | 1912–2021 | 110 |
| S10 | 1902–2021 | 120 |
| S11 | 1892–2021 | 130 |
| S12 | 1882–2021 | 140 |
| S13 | 1872–2021 | 150 |

### *3.1. Homogeneity Test Results*

The monthly precipitation data from the Durham meteorological station were first subjected to absolute homogeneity tests using the two-stage approach suggested by Wijngaard et al. [32]. Three methods were employed: Pettitt, Buishand, and SNHT. The results were checked for homogeneity regarding a 95% confidence level. Table 3 summarizes the main homogeneity results, including the tests' names, test statistics, and the homogeneity results. As a result of these tests, the precipitation data were found to be homogenous.

**Table 3.** Homogeneity test results for precipitation data.

| Test Name | Test Statistics | *p*-Value | Result | Homogeneity |
|---|---|---|---|---|
| Pettitt | 48,958 | 0.378 | Accept | Set 1: Homogenous |
| Buishand | 47.583 | 0.629 | Accept | |
| SNHT | 6.671 | 0.328 | Accept | |

### *3.2. Drought Characteristics for SPI 1*

The results of SPI 1 are interpreted and compared for thirteen different time period scenarios. Two drought events (Drought 1 (September 1996–October 1996) and Drought 2 (March 2011–June 2011)) are selected as examples and representative events, summarized in Table 4. Because SPI 1 is based on a 1-month time scale, the drought duration is generally low, ranging from 0 to 4 months. For the first drought event, increasing the time period above 110 years leads to no drought events (zero values), and the fifth, sixth, and seventh scenarios have no drought events. The duration in scenarios where the drought was recorded ranged consistently from 2 months upward across scenarios S1 through S4, S8, and S9, indicating a significant difference between these scenarios. The peak SPI values showed slight variability, ranging from −1.05 in Scenario S1 (30-year period) to −1 in Scenario S9 (110-year period), indicating a general uniformity in the peak drought experienced across these scenarios. In contrast, scenarios S5 through S13, which encompass longer historical records, did not record this drought event, possibly due to the changes in drought sensitivity over extended periods. The average SPI 1 (intensity) for the first drought event ranges between −0.73 and −0.78, showing a slight difference in intensities. However, the intensity for the second drought event ranges between −0.89 and −1.33, indicating a 33% difference; in this event, the maximum intensity is derived from S12 (140 years).

For Drought 2, a longer duration was observed, with all scenarios that recorded the drought showing durations from 2 to 4 months. The peak SPI values during this event were notably more severe, ranging from −2.41 in Scenario S1 to −2.35 in Scenario S13. This suggests a persistent severity in drought conditions when analyzed over a longer temporal frame. Notably, the severity, as measured by the sum of SPI, illustrated a marked increase in negative values. The median SPI values for drought events across all scenarios remained relatively consistent, underscoring the median as a robust measure of central drought tendency, less influenced by the extremities captured by the peak values. Because the number of drought months is low, the average and median values are relatively equal.

Figure 5 shows the variations in the drought index and characteristics for different time period scenarios (S1, S4, S7, S10, and S13). In Figure 5a, SPI 1 for S1 and S4 is less than −1, indicating the start of the drought event. However, for S7, S10, and S13, the SPI 1 values are more than −1, meaning no drought event has started. Figure 5b depicts the absolute values of peak SPI, intensity (average SPI), and median SPI for the abovementioned scenarios. Notably, the peak SPI values are relatively uniform across the scenarios, suggesting that the most intense phase of the drought is captured similarly regardless of the period. Figure 5c shows the duration of the drought alongside the absolute summation of SPI values, providing a measure of the cumulative severity of the drought across different scenarios. Based on the comprehensive analysis of all scenarios for the first drought event, the critical drought characteristics have been identified as an intensity of −0.78, a duration of 2 months, a peak SPI of −1.05, and a severity of SPI of −1.56.

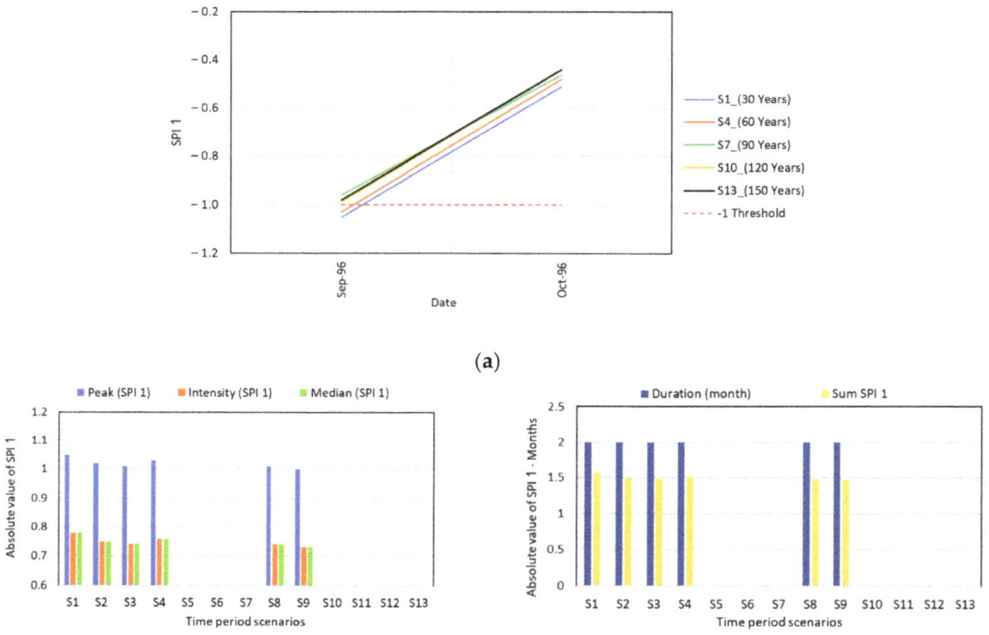


**Figure 5.** Potential drought characteristics for SPI 1 for the first selected drought event (September 1996–October 1996). (**a**) The difference between selected scenarios based on SPI 1, (**b**) the difference between all time period scenarios using the peak, intensity, and median values, (**c**) the difference between all time period scenarios using drought duration and the absolute summation of SPI 1 values.

Figure 6 presents the analysis of the potential drought characteristics for the second selected drought event (March 2011–June 2011) using the SPI. Figure 6a depicts the SPI values over the drought period for selected scenarios (S1, S4, S7, S10, and S13), highlighting consistent behavior but differentiating in the starting SPI index value. Figure 6b illustrates the absolute values of peak SPI, intensity (average SPI), and median SPI for each scenario. The peak SPI, representing the most severe point in the drought index, shows higher values for shorter time period scenarios (S1, S4) and slightly less severity in longer time frames (S10, S13), indicating that longer records have lesser peak values due to the longer climatic periods in the analysis. Conversely, the intensity and median values offer a clearer view of how the drought persists over its duration, with significant differences (Figure 6c). Based on the comprehensive analysis of all scenarios for the second drought event, the critical drought characteristics have been identified as an intensity of −1.33, a duration of 4 months, a peak SPI of −2.43, and a severity of the SPI of −3.9.

**Table 4.** Summary of the drought characteristics for SPI 1 for different 13 time period scenarios.

| Drought Characteristics SPI 1 | | | | | | | | | | | |
|---|---|---|---|---|---|---|---|---|---|---|---|
| Years | | Drought 1 (September 1996–October 1996) | | | | | Drought 2 (March 2011–June 2011) | | | | |
| Scenario | | D | P | S | A | M | D | P | S | A | M |
| S1—30 Y | 1992–2021 | 2 | −1.05 | −1.56 | −0.78 | −0.78 | 4 | −2.41 | −3.9 | −0.97 | −0.71 |
| S2—40 Y | 1982–2021 | 2 | −1.02 | −1.49 | −0.75 | −0.75 | 4 | −2.39 | −3.77 | −0.94 | −0.67 |
| S3—50 Y | 1972–2021 | 2 | −1.01 | −1.48 | −0.74 | −0.74 | 3 | −2.39 | −2.73 | −0.91 | −0.32 |
| S4—60 Y | 1962–2021 | 2 | −1.03 | −1.51 | −0.76 | −0.76 | 4 | −2.43 | −3.82 | −0.95 | −0.68 |
| S5—70 Y | 1952–2021 | 0 | - | - | - | - | 3 | −2.36 | −2.67 | −0.89 | −0.30 |
| S6—80 Y | 1942–2021 | 0 | - | - | - | - | 3 | −2.36 | −2.66 | −0.89 | −0.30 |
| S7—90 Y | 1932–2021 | 0 | - | - | - | - | 3 | −2.37 | −2.71 | −0.90 | −0.32 |
| S8—100 Y | 1922–2021 | 2 | −1.01 | −1.48 | −0.74 | −0.74 | 4 | −2.39 | −3.74 | −0.94 | −0.66 |
| S9—110 Y | 1912–2021 | 2 | −1 | −1.47 | −0.73 | −0.73 | 3 | −2.38 | −2.72 | −0.91 | −0.32 |
| S10—120 Y | 1902–2021 | 0 | - | - | - | - | 3 | −2.36 | −2.67 | −0.89 | −0.30 |
| S11—130 Y | 1892–2021 | 0 | - | - | - | - | 2 | −2.37 | −2.66 | −1.33 | −1.33 |
| S12—140 Y | 1882–2021 | 0 | - | - | - | - | 2 | −2.35 | −2.64 | −1.32 | −1.32 |
| S13—150 Y | 1872–2021 | 0 | - | - | - | - | 3 | −2.36 | −2.66 | −0.89 | −0.89 |

D: duration, P: peak SPI, S: sum of SPI, A: average SPI (intensity), M: median SPI.

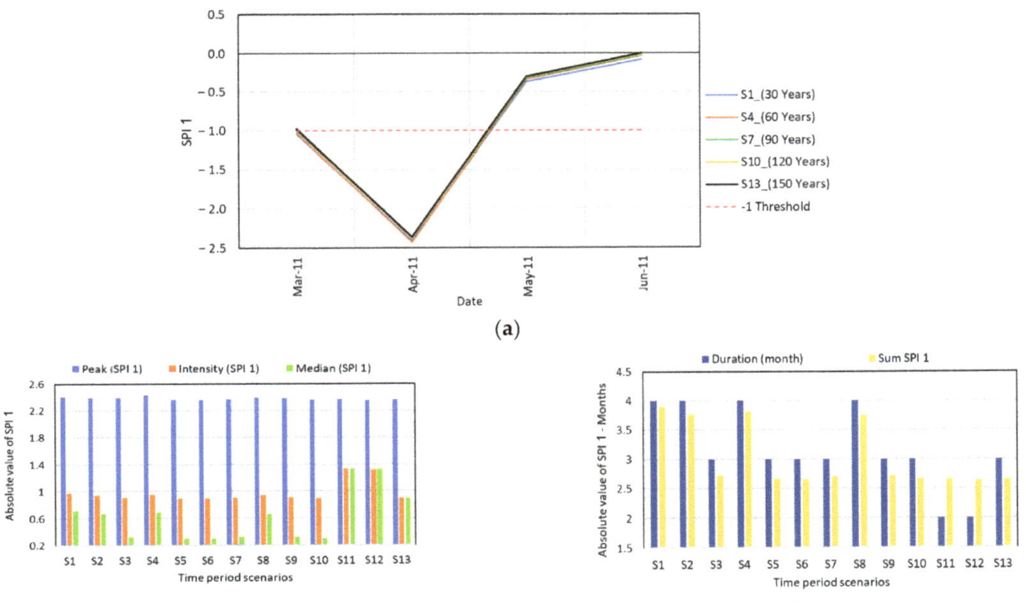


**Figure 6.** Potential drought characteristics for SPI 1 for the second selected drought event (March 2011–June 2011). (**a**) The difference between selected scenarios based on SPI 1, (**b**) the difference between all time period scenarios using the peak, intensity, and median values, (**c**) the difference between all time period scenarios using drought duration and the absolute summation of SPI 1 values.

### 3.3. Drought Characteristics for SPI 3

The results of SPI 3 are conducted on and compared for thirteen different time period scenarios. Two drought events (Drought 1 (November 2011–April 2012) and Drought 2 (July 2018–May 2019)) are selected as representative examples and events, summarized in Table 5 and presented in Figures 7 and 8. For Drought 1, the durations of the drought events varied between the scenarios, with longer durations observed in the 30-year, 40-year, and 60-year scenarios (S1, S2, and S4), at 6 months each, while the other scenarios found drought durations of 2 months. The peak SPI 3 values ranged from −2.55 in S1 to −2.42 in S6, indicating severe drought conditions across all scenarios. The sum of SPI values, representing the severity, also showed significant negative values, underscoring the intense nature of this drought event (first drought event). Notably, the absolute values of the median SPI and SPI intensity varied slightly, suggesting a consistent median drought condition across different historical time period scenarios but variable intensities, which were slightly more severe in longer period scenarios. The SPI intensity ranged between −1.08 (S2) and −1.91 (S12), indicating a 50% increase in intensity. Based on these results, the critical drought characteristics have been identified as an intensity of −1.91, a duration of 6 months, a peak SPI of −2.55, and a severity SPI of −6.74.

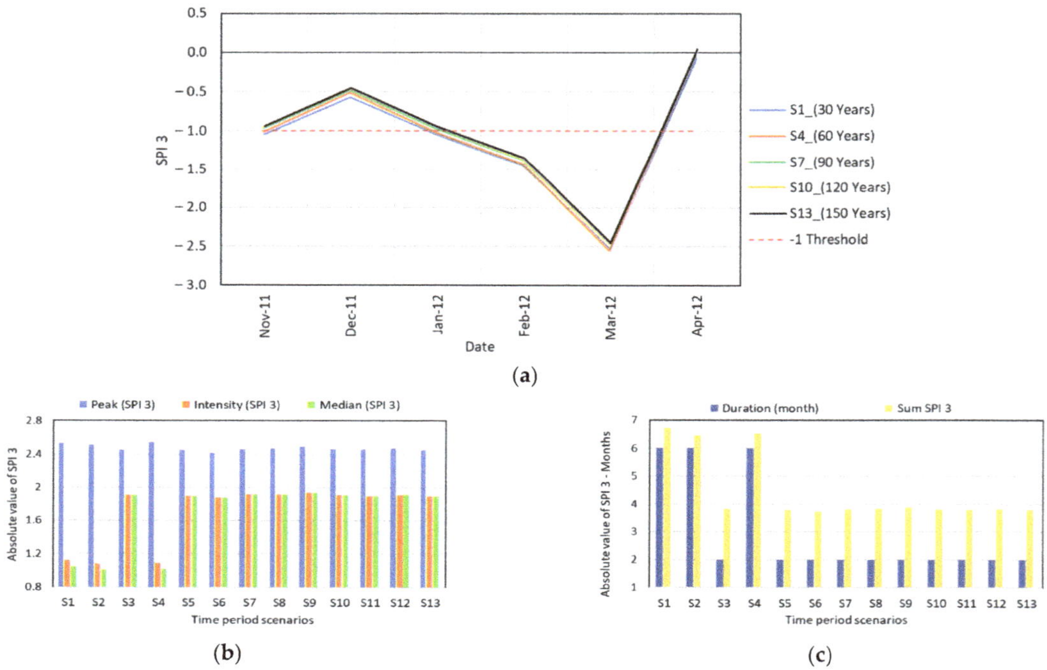


**Figure 7.** Potential drought characteristics for SPI 3 for the first selected drought event (November 2011–April 2012). (**a**) The difference between selected scenarios based on SPI 3, (**b**) the difference between all time period scenarios using the peak, intensity, and median values, (**c**) the difference between all time period scenarios using drought duration and the absolute summation of SPI 3 values.

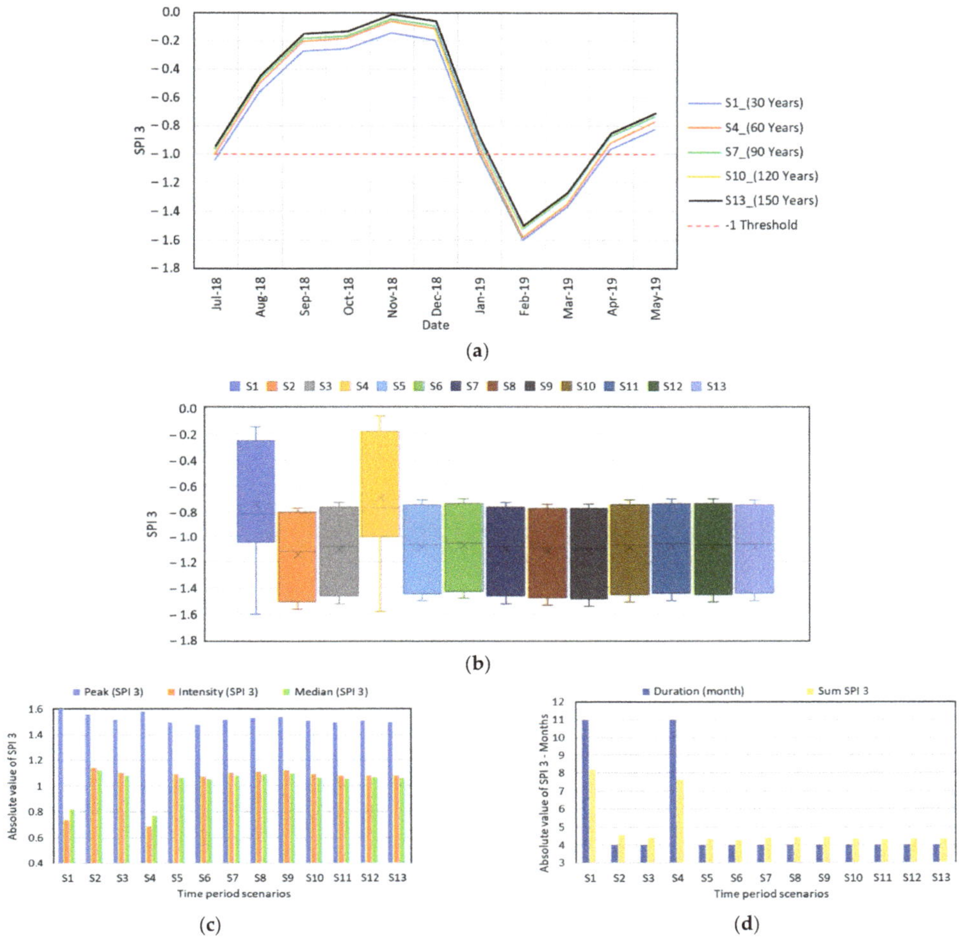


**Figure 8.** Potential drought characteristics for SPI 3 for the second selected drought event (July 2018–May 2019). (**a**) The difference between selected scenarios based on SPI 3, (**b**) the box-and-whisker plot for SPI 3 values, (**c**) the difference between all time period scenarios using the peak, intensity, and median values, (**d**) the difference between all time period scenarios using drought duration and the absolute summation of SPI 3 values.

Figure 8 presents a detailed analysis of the potential drought characteristics for SPI 3 during the second selected drought event (July 2018–May 2019) across multiple time period scenarios. Figure 8a shows the SPI 3 values for five selected scenarios (30, 60, 90, 120, and 150 years), where S1, S2, and S4 fell below −1 in July 2018, and the other scenarios fell below −1 in February 2019. Figure 8b, a box-and-whisker plot, displays the distribution of SPI values across all scenarios during the drought event, highlighting variability in the central tendency and SPI values. Figure 8c shows each scenario's peak, intensity, and median SPI value, illustrating how these characteristics differ across various historical time period scenarios. Figure 8d compares the duration and severity of the drought across all scenarios, revealing that while the duration of the drought is relatively consistent, the total severity varies, with shorter scenarios tending to accumulate higher negative SPI values, leading to more intense drought conditions. For S1, S2, and S4, the duration is 11 months. Based on these results, the critical drought characteristics have been identified as an intensity of −1.14, a duration of 11 months, a peak SPI of −1.60, and a severity SPI of −8.19.

**Table 5.** Summary of the drought characteristics for SPI 3 for different 13 time period scenarios.

| Drought Characteristics SPI 3 | | | | | | | | | | | |
|---|---|---|---|---|---|---|---|---|---|---|---|
| Years | | Drought 1 (November 2011–April 2012) | | | | | Drought 2 (July 2018–May 2019) | | | | |
| Scenario | | D | P | S | A | M | D | P | S | A | M |
| S1—30 Y | 1992–2021 | 6 | −2.53 | −6.74 | −1.12 | −1.05 | 11 | −1.60 | −8.19 | −0.74 | −0.82 |
| S2—40 Y | 1982–2021 | 6 | −2.51 | −6.46 | −1.08 | −1.01 | 4 | −1.56 | −4.57 | −1.14 | −1.12 |
| S3—50 Y | 1972–2021 | 2 | −2.45 | −3.82 | −1.91 | −1.91 | 4 | −1.52 | −4.41 | −1.1 | −1.08 |
| S4—60 Y | 1962–2021 | 6 | −2.55 | −6.53 | −1.09 | −1.02 | 11 | −1.58 | −7.62 | −0.69 | −0.77 |
| S5—70 Y | 1952–2021 | 2 | −2.45 | −3.8 | −1.9 | −1.9 | 4 | −1.50 | −4.35 | −1.09 | −1.07 |
| S6—80 Y | 1942–2021 | 2 | −2.42 | −3.75 | −1.88 | −1.88 | 4 | −1.48 | −4.28 | −1.07 | −1.05 |
| S7—90 Y | 1932–2021 | 2 | −2.46 | −3.83 | −1.92 | −1.92 | 4 | −1.52 | −4.42 | −1.1 | −1.08 |
| S8—100 Y | 1922–2021 | 2 | −2.47 | −3.84 | −1.92 | −1.92 | 4 | −1.53 | −4.44 | −1.11 | −1.09 |
| S9—110 Y | 1912–2021 | 2 | −2.49 | −3.88 | −1.94 | −1.94 | 4 | −1.54 | −4.47 | −1.12 | −1.10 |
| S10—120 Y | 1902–2021 | 2 | −2.46 | −3.82 | −1.91 | −1.91 | 4 | −1.51 | −4.35 | −1.09 | −1.07 |
| S11—130 Y | 1892–2021 | 2 | −2.46 | −3.81 | −1.9 | −1.9 | 4 | −1.5 | −4.31 | −1.08 | −1.06 |
| S12—140 Y | 1882–2021 | 2 | −2.47 | −3.83 | −1.91 | −1.91 | 4 | −1.51 | −4.34 | −1.08 | −1.07 |
| S13—150 Y | 1872–2021 | 2 | −2.45 | −3.81 | −1.9 | −1.9 | 4 | −1.5 | −4.34 | −1.08 | −1.06 |

D: duration, P: peak SPI, S: sum of SPI, A: average SPI (intensity), M: median SPI.

### *3.4. Drought Characteristics for SPI 6*

For a 6-month time scale, two drought events were used to compare the results obtained for thirteen scenarios using 150 years of precipitation data from Durham station. The first drought event occurred in 1991–1992 and the second in 2017. Drought indices and detailed drought characteristics obtained from the classical SPI analysis for each scenario are summarized in Table 6 and Figures 9 and 10. For the first drought event, the durations varied significantly across scenarios, with longer durations generally observed in longer scenarios (40 to 150 years), where drought lasted 12 to 13 months. This contrasts with the shortest scenario (S1, 30 years), which recorded a duration of only 3 months. The peak SPI values during this event were notably severe across all scenarios, ranging from −2.53 in S1 to −1.86 in S6, highlighting that this drought event was consistently captured as intense across different time period scenarios. The maximum intensity was observed in S2 (40 years) with a value of −1 and 13 months of drought.

**Table 6.** Summary of the drought characteristics for SPI 6 for different 13 time period scenarios.

| Drought Characteristics SPI 6 | | | | | | | | | | | |
|---|---|---|---|---|---|---|---|---|---|---|---|
| Years | | Drought 1 (August 1991–August 1992) | | | | | Drought 2 (May 2017–August 2017) | | | | |
| Scenario | | D | P | S | A | M | D | P | S | A | M |
| S1—30 Y | 1992–2021 | 3 | −1.08 | −1.94 | −0.65 | −0.65 | 4 | −1.42 | −2.44 | −0.61 | −0.45 |
| S2—40 Y | 1982–2021 | 13 | −2.02 | −13.06 | −1 | −1 | 4 | −1.35 | −2.07 | −0.52 | −0.35 |
| S3—50 Y | 1972–2021 | 13 | −1.91 | −12.08 | −0.93 | −0.93 | 2 | −1.26 | −1.72 | −0.86 | −0.86 |
| S4—60 Y | 1962–2021 | 13 | −1.98 | −12.69 | −0.98 | −0.97 | 4 | −1.32 | −1.96 | −0.49 | −0.32 |
| S5—70 Y | 1952–2021 | 13 | −1.90 | −11.86 | −0.91 | −0.91 | 2 | −1.25 | −1.69 | −0.84 | −0.84 |
| S6—80 Y | 1942–2021 | 13 | −1.86 | −11.59 | −0.89 | −0.89 | 2 | −1.22 | −1.65 | −0.82 | −0.82 |
| S7—90 Y | 1932–2021 | 13 | −1.93 | −12.19 | −0.94 | −0.94 | 2 | −1.27 | −1.74 | −0.87 | −0.87 |
| S8—100 Y | 1922–2021 | 13 | −1.94 | −12.31 | −0.95 | −0.94 | 2 | −1.28 | −1.76 | −0.88 | −0.88 |
| S9—110 Y | 1912–2021 | 13 | −1.96 | −12.37 | −0.95 | −0.95 | 2 | −1.29 | −1.76 | −0.88 | −0.88 |
| S10—120 Y | 1902–2021 | 13 | −1.92 | −11.84 | −0.91 | −0.91 | 2 | −1.25 | −1.68 | −0.84 | −0.84 |
| S11—130 Y | 1892–2021 | 12 | −1.90 | −11.62 | −0.97 | −0.96 | 2 | −1.23 | −1.65 | −0.82 | −0.82 |
| S12—140 Y | 1882–2021 | 12 | −1.91 | −11.68 | −0.97 | −0.97 | 2 | −1.24 | −1.66 | −0.83 | −0.83 |
| S13—150 Y | 1872–2021 | 13 | −1.89 | −11.65 | −0.90 | −0.96 | 2 | −1.23 | −1.65 | −0.83 | −0.83 |

D: duration, P: peak SPI, S: sum of SPI, A: average SPI (intensity), M: median SPI.

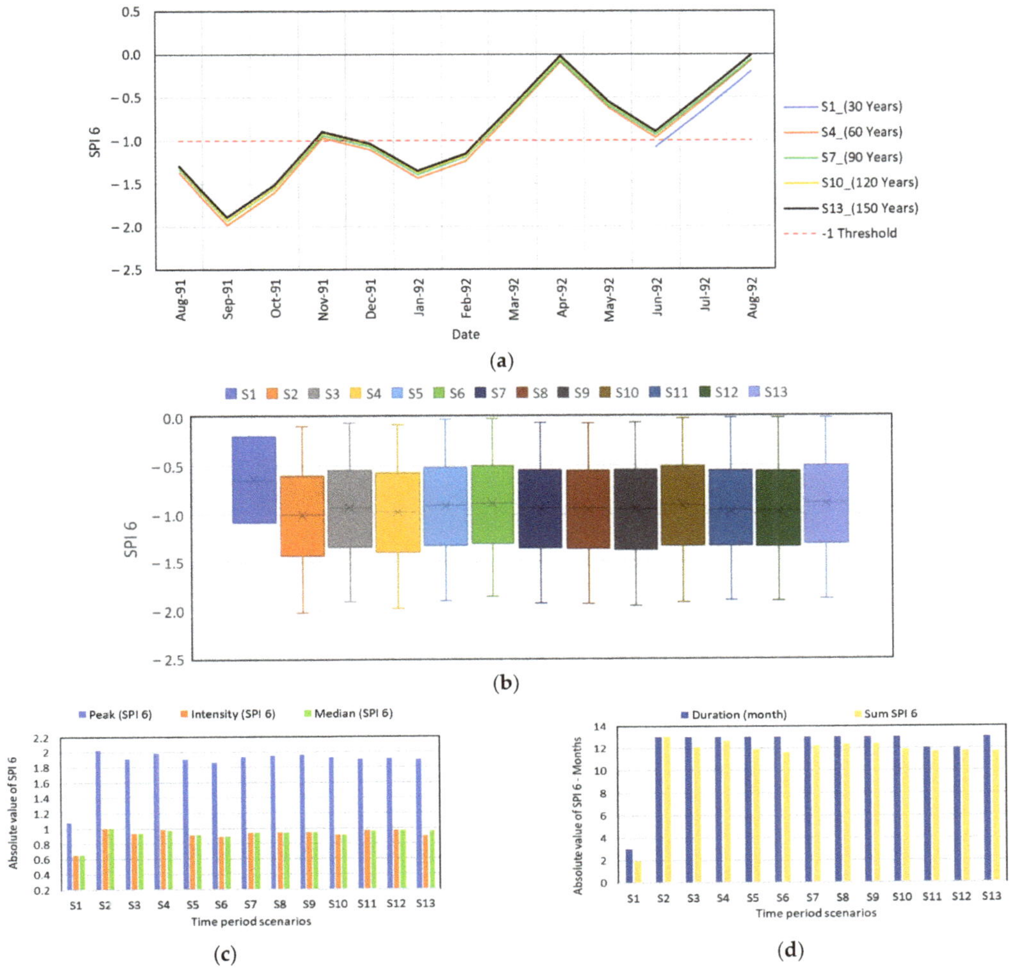


**Figure 9.** Potential drought characteristics for SPI 6 for the first selected drought event (August 1991–August 1992). (**a**) The difference between selected scenarios based on SPI 6, (**b**) the box-and-whisker plot for SPI 6 values, (**c**) the difference between all time period scenarios using the peak, intensity, and median values, (**d**) the difference between all time period scenarios using drought duration and the absolute summation of SPI 6 values.

In Figure 9a, the SPI values for selected scenarios showed that all the scenarios went below the −1 threshold during the drought event, except the first scenario, which started in June 1992. The box-and-whisker plot in Figure 9b provides a distribution of the SPI values during this drought event across all scenarios, illustrating the variability in the drought index values captured in different time periods. This visualization highlights that while the median SPI values are fairly consistent, the range of peak and average values can vary, reflecting different perceptions of drought analysis based on the different time period scenarios. The bar charts in Figure 9c,d depict the absolute values of peak SPI, intensity, median, and drought duration and severity. Based on these results, the critical drought characteristics have been identified as an intensity of −1, a duration of 13 months, a peak SPI of −2.02, and a severity SPI of −13.06.

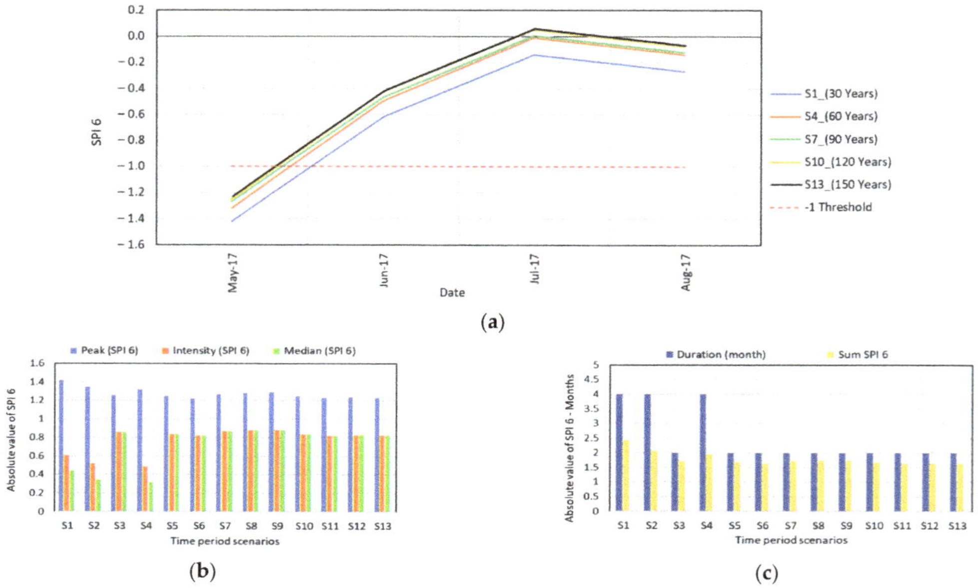


**Figure 10.** Potential drought characteristics for SPI 6 for the second selected drought event (May 2017–August 2017). (**a**) The difference between selected scenarios based on SPI 6, (**b**) the difference between all time period scenarios using the peak, intensity, and median values, (**c**) the difference between all time period scenarios using drought duration and the absolute summation of SPI 6 values.

Figure 10 presents the analysis of potential drought characteristics for the second selected drought event (May 2017–August 2017). Figure 10a depicts the SPI values over the drought period for selected scenarios, highlighting a difference in duration between these scenarios. For S1 and S4, the duration is 4 months; for other scenarios, the duration is 2. However, all the scenarios started in the same month, but the SPI value in the third and fourth months was the main reason for the change in the duration. Figure 10b illustrates the absolute peak SPI, intensity, and median values for each scenario. Figure 10c shows the duration and the absolute values of the severity. Based on the comprehensive analysis of all scenarios for the second drought event, the critical drought characteristics have been identified as an intensity of −0.88, a duration of 4 months, a peak SPI of −1.42, and a severity of SPI of −2.44.

### 3.5. Drought Characteristics for SPI 12

This section explains the findings from an analysis of SPI 12 across 13 different time period scenarios for a prolonged drought event that occurred from August 1995 to December 1997, as summarized in Table 7 and shown in Figure 11. The analysis reveals considerable variations in the drought's characteristics based on the time period scenarios used. The duration of the drought event ranged widely, with the shortest period (S1, 30 years) experiencing the longest drought duration at 29 months, while most other scenarios had durations of about 20 to 23 months. This variability in duration highlights how the selected time period can influence the drought characteristics. Also, the intensity ranged between −0.93 and −1.09. The maximum duration, severity, and intensity were observed in S1.

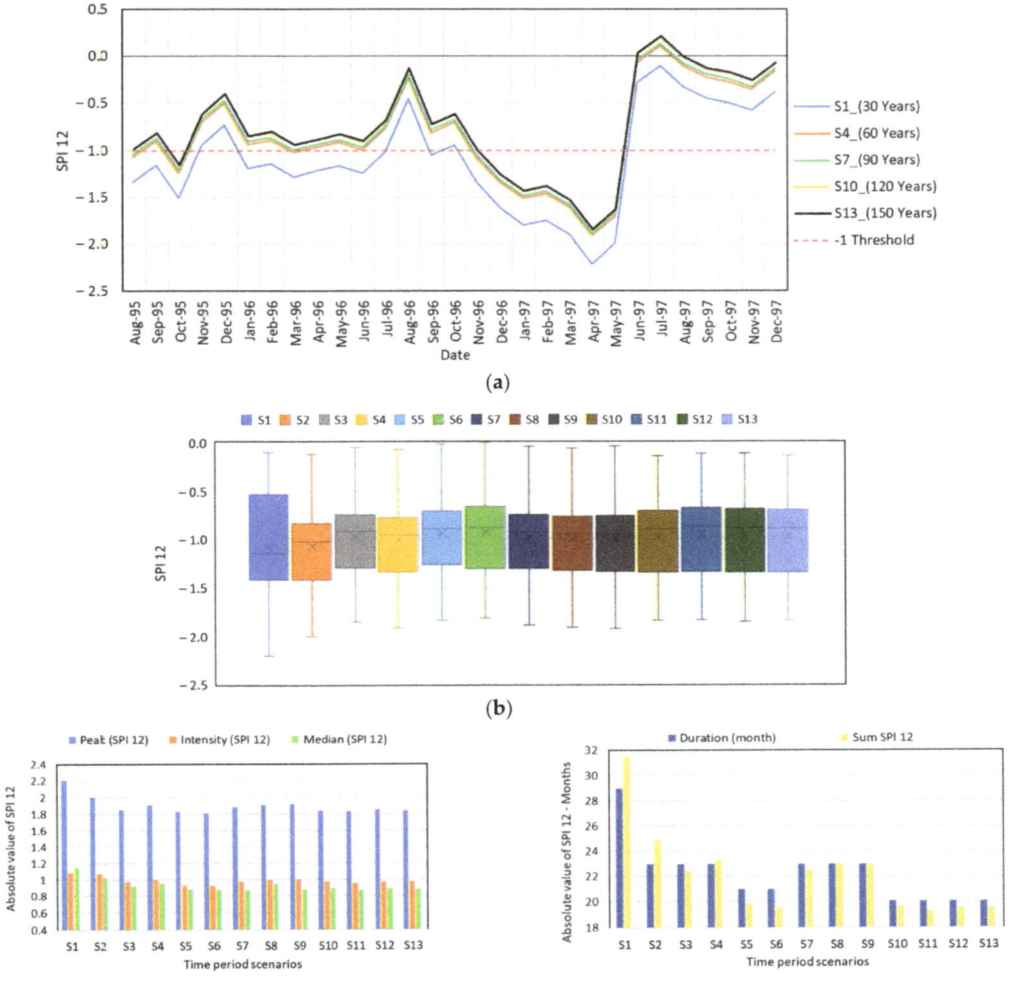


**Figure 11.** Potential drought characteristics for SPI 12 for the first selected drought event (August 1995–December 1997). (**a**) The difference between selected scenarios based on SPI 12, (**b**) the box-and-whisker plot for SPI 12 values, (**c**) the difference between all time period scenarios using the peak, intensity, and median values, (**d**) the difference between all time period scenarios using drought duration and the absolute summation of SPI 12 values.

Figure 11a shows that while the SPI values fluctuated, the start and end years differed for the time period scenarios. For example, S4, S7, S10, and S13 ended in July 1997. In contrast, S1 ended in January 1998. The box-and-whisker plot in Figure 11b visualizes the distribution of SPI values for each scenario during the drought. Figure 11c quantifies the peak, intensity, and median SPI values for each time period scenario, indicating some consistency in these metrics across the scenarios. Figure 11d compares the duration and severity of the drought across all scenarios. There is a noticeable difference in how the duration of droughts and severity are perceived, with shorter scenarios tending to record longer durations and a higher cumulative severity. Based on the comprehensive analysis of all scenarios for the second drought event, the critical drought characteristics have been identified as an intensity of −1.09, a duration of 29 months, a peak SPI of −2.21, and a severity of SPI of −31.48.

**Table 7.** Summary of the drought characteristics for SPI 12 for different 13 time period scenarios.

| Drought Characteristics SPI 12 | | | | | | |
|---|---|---|---|---|---|---|
| Years | | Drought 1 (August 1995–December 1997) | | | | |
| Scenario | | D | P | S | A | M |
| S1—30 Y | 1992–2021 | 29 | −2.21 | −31.48 | −1.09 | −1.15 |
| S2—40 Y | 1982–2021 | 23 | −2 | −24.89 | −1.08 | −1.03 |
| S3—50 Y | 1972–2021 | 23 | −1.85 | −22.43 | −0.98 | −0.92 |
| S4—60 Y | 1962–2021 | 23 | −1.91 | −23.3 | −1.01 | −0.96 |
| S5—70 Y | 1952–2021 | 21 | −1.83 | −19.82 | −0.94 | −0.89 |
| S6—80 Y | 1942–2021 | 21 | −1.81 | −19.53 | −0.93 | −0.88 |
| S7—90 Y | 1932–2021 | 23 | −1.88 | −22.56 | −0.98 | −0.87 |
| S8—100 Y | 1922–2021 | 23 | −1.91 | −23.06 | −1 | −0.95 |
| S9—110 Y | 1912–2021 | 23 | −1.92 | −22.99 | −1 | −0.88 |
| S10—120 Y | 1902–2021 | 20 | −1.84 | −19.66 | −0.98 | −0.90 |
| S11—130 Y | 1892–2021 | 20 | −1.83 | −19.26 | −0.96 | −0.88 |
| S12—140 Y | 1882–2021 | 20 | −1.85 | −19.49 | −0.97 | −0.89 |
| S13—150 Y | 1872–2021 | 20 | −1.84 | −19.57 | −0.98 | −0.89 |

D: duration, P: peak SPI, S: sum of SPI, A: average SPI (intensity), M: median SPI.

## 4. Discussion

### *4.1. Initial Time Condition and Dynamic Time Period Scenarios*

The determination of the initial time conditions and dynamic time period scenarios plays a pivotal role in calculating the drought characteristics, particularly in critical drought assessments. This step, as a first step, is crucial, as the results heavily depend on it. Dynamic time period scenarios are established based on the available data, the specific purpose of the drought characteristics analysis, and the climate conditions of the study area. This study utilized 10-year intervals, resulting in 13 time period scenarios. However, for more precise and localized studies, shorter intervals, such as one year, may be necessary. Therefore, the selection of dynamic time period scenarios should be tailored to both the available data and the objectives of the drought characteristics analysis. For example, for more accurate and microscale studies, smaller intervals are recommended. Additionally, in arid regions where drought events have significant impacts, smaller intervals are recommended to capture finer-scale variations in drought severity and duration.

### *4.2. Drought Definition and Critical Drought Characteristics*

The definition of drought and its relationship to drought characteristics is a fundamental aspect of drought analysis. In this study, we used the drought definition proposed by the original article on SPI theory [16], which sets the threshold for drought initiation at an SPI value of −1. However, it is important to recognize that alternative definitions, such as those based on run theory [35], or different definitions such as −1.0 or 0 [37], can lead to different results. This highlights the need for further research to understand the impact of the selected drought definition and threshold on drought characteristics and critical drought assessments. McKee et al. [16] emphasized the use of a threshold of −1, arguing that values between 0 and −1 may still reflect normal or wet conditions and thus do not signify the onset of a drought event. This choice is pivotal, as the SPI provides a versatile and universally applicable method that is adaptable to different time scales and sensitive to changes in precipitation patterns [38,39]. Furthermore, employing a threshold of −1 instead of 0 results in shorter drought durations and a higher intensity, providing a more conservative approach to identifying critical drought characteristics. Clarifying the implications of different drought definitions and thresholds is essential for enhancing the accuracy and reliability of drought assessments and management strategies.

### 4.3. Comparison between Critical and Traditional Drought Characteristics

In terms of the drought characteristics obtained in this research, the analysis reveals noteworthy insights across dynamic time period scenarios and SPI time scales. For SPI 1, the intensity ranges between −0.89 and −1.33, indicating a 33% difference, with the maximum intensity derived from the longest time period (S12, 140 years). This suggests that increasing the time period for short time scales yields more critical and conservative results. Additionally, variations in drought duration are observed, with some scenarios showing no drought events, particularly in longer time period scenarios, implying that increasing the time period decreases the duration. Conversely, for SPI 3, there is a positive relationship between time period and drought intensity, while duration exhibits an inverse relationship, similar to SPI 1, where shorter time periods yield longer durations. Decreasing the time period leads to longer durations, consequently resulting in reduced intensity, as intensity is inversely proportional to severity divided by drought duration. However, for SPI 6, no consistent relationship is observed between duration and intensity, necessitating careful consideration, especially for short time scales. In contrast, for longer time scales like SPI 12, an adverse relationship between time period and intensity and duration is observed, with increasing time periods resulting in decreased intensity and duration. Therefore, utilizing shorter time periods, such as 30 years, is deemed more conservative. For instance, the duration decreases from 29 months for a 30-year time period to 20 months for a 150-year time period, while the intensity decreases from −1.09 to −0.98. This trend may be attributed to the increased impact of climate change over the last 30 years. These findings underscore the importance of considering both the time period and SPI time scales in drought analysis for effective drought management and decision-making.

### 4.4. Critical Drought Characteristics and Various Sectors

Utilizing critical drought characteristics, which offer a more nuanced and precise understanding of drought characteristics and impacts compared to traditional drought characteristics, can significantly enhance decision-making and design across various sectors. For instance, in water resource management, critical drought characteristics provide insights into the most severe and prolonged drought events, enabling more effective allocation of water resources and infrastructure planning [40,41]. For example, reservoir design and operation plans may incorporate critical drought characteristics to ensure a sufficient water supply during prolonged dry periods. Agricultural stakeholders can benefit from critical drought characteristics by identifying specific crop water requirements and implementing targeted irrigation strategies to mitigate the impacts of extreme drought conditions [42]. Moreover, critical drought characteristics are invaluable in ecosystem management, guiding conservation efforts to protect vulnerable ecosystems and biodiversity from the adverse effects of severe droughts. By incorporating critical drought characteristics into decision-making processes and design considerations, stakeholders can enhance resilience to drought and better adapt to the challenges posed by water scarcity.

### 4.5. Previous Studies

Previous studies, such as those conducted by Wang et al. [43] and Laimighofer and Laaha [44], have extensively investigated drought uncertainties, including the effect of the selected time periods. Both studies underscored the observation period as one of the most significant sources of uncertainty in drought analysis. However, despite acknowledging this, they primarily focused on drought index values without delving into analysis or quantification of the observation period's impact on the drought characteristics and their corresponding applications. Furthermore, these studies did not provide any perspective or suggestions on determining the critical drought conditions arising from these uncertainties. For instance, Laimighofer and Laaha [44] highlighted that the observation period can account for up to 49% of the uncertainty in SPI calculations. Conversely, Wang et al. [43] stated that the uncertainty in SPI calculation decreases with an increase in the time scale and record length. These findings align with previous literature. Also, in terms of critical drought characteristics and their corresponding applications, there is no universally preferable time period. Instead, each drought event and its critical characteristics have an event-based critical time period. However, our research delved further, aiming to uncover the nuanced effects of the time period on drought characteristics, particularly focusing on identifying critical drought characteristics essential for effective drought management and adaptation strategies.

### *4.6. New Parameters for Drought Characterization*

This study extends beyond the classical drought characteristics traditionally used in drought analysis, such as duration, severity, and intensity, to incorporate additional parameters that offer a more accurate understanding of drought dynamics. Drought characteristics offer detailed insights into droughts' temporal and quantitative aspects, allowing for a nuanced understanding of their progression and mitigation [35–37]. Specifically, this research calculates the peak and median values of the drought index during each identified drought event and scenario. The peak value represents the maximum drought index, providing insight into the most severe point of precipitation deficiency experienced during the event. This metric is crucial for understanding the potential stress on ecological and agricultural systems. Similarly, the median value of the drought index serves as a robust measure of the typical drought conditions over the event's duration, offering a more stable indicator that is less influenced by extreme values than the mean. This helps in ensuring that the assessment is not skewed by unusually wet or dry values within the drought period.

### *4.7. Limitations and Future Opportunities*

Like any research, this study has certain limitations. One significant limitation is the use of data from only a single meteorological station, chosen for its long and continuous data records. The inclusion of additional stations might yield new insights, potentially enhancing the robustness and generalizability of the findings. However, given that this article proposes new concepts and methods regarding selecting the time periods, temporal drought evaluation, and critical drought characteristics, an application to one station is sufficient. One station is enough to demonstrate the effectiveness and applicability of the proposed methodologies. Another limitation is the reliance on a single drought definition, specifically the one introduced by McKee et al. [16]. Future research could benefit from exploring various drought definitions and different time period scenarios to provide a more comprehensive understanding of critical drought characteristics. By addressing these limitations, subsequent studies can build on this work to refine and expand the applicability of the proposed methods.

## 5. Conclusions

This research significantly advances the field of drought analysis by examining the impact of various time period scenarios on drought characteristics and introducing a novel methodology that employs dynamic time periods to identify critical drought characteristics. Through the analysis of drought indices across different time periods, this study underscores the importance of temporal variability in understanding drought and provides a novel approach to filling this gap. The proposed methodology enhances the precision of identifying critical drought characteristics and offers valuable insights for water resource management, drought mitigation planning, and infrastructure design. This study lays a solid foundation for future work to improve the accuracy and adaptability of drought assessments and management strategies. The key findings can be summarized as follows:

1. Significant differences in drought characteristics were observed across different time period scenarios.
2. The duration of drought events varied notably when different time periods were considered. For example, for SPI 12, the drought duration varied significantly from 20 to 29 months, and for SPI 6, the drought duration varied between 3 and 13 months.

3. The intensity of SPI 1 ranged between −0.89 and −1.33, indicating a 33% increase, and the SPI 3 intensity ranged between −1.08 and −1.91, indicating a 50% increase.
4. The proposed methodology using dynamic time period scenarios instead of one time period enhances the precision of identifying critical drought characteristics.
5. The selection of a definition of droughts significantly impacts the resulting drought characteristics, highlighting the need for careful selection and further research to understand the implications of different definitions on drought assessments.

**Author Contributions:** Conceptualization, A.A.A. and E.Ş.; methodology, A.A.A. and E.Ş.; software, A.A.A.; validation, E.Ş., M.E.B. and Ş.A.G.; formal analysis, A.A.A. and E.Ş.; investigation, A.A.A. and E.Ş.; resources, E.Ş., Ş.A.G. and M.E.B.; data curation, E.Ş.; writing—original draft preparation, A.A.A.; writing—review and editing, E.Ş., Ş.A.G. and M.E.B.; visualization, A.A.A. and E.Ş.; supervision, E.Ş. and M.E.B.; project administration, E.Ş., Ş.A.G. and M.E.B.; funding acquisition, E.Ş. and Ş.A.G. All authors have read and agreed to the published version of the manuscript.

**Funding:** This research received no external funding.

**Institutional Review Board Statement:** Not applicable.

**Informed Consent Statement:** Not applicable.

**Data Availability Statement:** The precipitation data are available at https://durhamweather.webspace.durham.ac.uk/ (accessed on 25 October 2023).

**Acknowledgments:** The authors would like to acknowledge that this paper is submitted in partial fulfillment of the requirements for the Ph.D. degree at Yildiz Technical University.

**Conflicts of Interest:** The authors declare no conflicts of interest.

## References

1. Muhammad, W.; Yang, H.; Lei, H.; Muhammad, A.; Yang, D. Improving the regional applicability of satellite precipitation products by ensemble algorithm. *Remote Sens.* **2018**, *10*, 577. [CrossRef]
2. Kesgin, E.; Yaldız, S.G.; Güçlü, Y.S. Spatiotemporal variability and trends of droughts in the Mediterranean coastal region of Türkiye. *Int. J. Climatol.* **2024**, *44*, 1036–1057. [CrossRef]
3. Kharyutkina, E.; Loginov, S.; Martynova, Y.; Sudakov, I. Time series analysis of atmospheric precipitation characteristics in Western Siberia for 1979–2018 across different datasets. *Atmosphere* **2022**, *13*, 189. [CrossRef]
4. Vaheddoost, B.; Mohammadi, B.; Safari, M.J.S. The Association between Meteorological Drought and the State of the Groundwater Level in Bursa, Turkey. *Sustainability* **2023**, *15*, 15675. [CrossRef]
5. Merabti, A.; Meddi, M.; Martins, D.S.; Pereira, L.S. Comparing SPI and RDI applied at local scale as influenced by climate. *Water Resour. Manag.* **2018**, *32*, 1071–1085. [CrossRef]
6. Nassaj, B.N.; Zohrabi, N.; Shahbazi, A.N.; Fathian, H. Evaluating the performance of eight global gridded precipitation datasets across Iran. *Dyn. Atmos. Ocean.* **2022**, *98*, 101297. [CrossRef]
7. IPCC. Summary for policymakers. In *Climate Change 2023: Synthesis Report. Contribution of Working Groups I, II and III to the Sixth Assessment Report of the Intergovernmental Panel on Climate Change*; Core Writing Team, Lee, H., Romero, J., Eds.; IPCC: Geneva, Switzerland, 2023; pp. 1–34. [CrossRef]
8. Zhao, X.; Xia, H.; Liu, B.; Jiao, W. Spatiotemporal comparison of drought in Shaanxi–Gansu–Ningxia from 2003 to 2020 using various drought indices in google earth engine. *Remote Sens.* **2022**, *14*, 1570. [CrossRef]
9. Gu, L.; Chen, J.; Yin, J.; Sullivan, S.C.; Wang, H.M.; Guo, S.; Zhang, L.; Kim, J.S. Projected increases in magnitude and socioeconomic exposure of global droughts in 1.5 and 2 °C warmer climates. *Hydrol. Earth Syst. Sci.* **2020**, *24*, 451–472. [CrossRef]
10. Zhao, Q.; Chen, Q.; Jiao, M.; Wu, P.; Gao, X.; Ma, M.; Hong, Y. The temporal-spatial characteristics of drought in the Loess Plateau using the remote-sensed TRMM precipitation data from 1998 to 2014. *Remote Sens.* **2018**, *10*, 838. [CrossRef]
11. Brito, C.S.; da Silva, R.M.; Santos, C.A.G.; Neto, R.M.B.; Coelho, V.H.R. Monitoring meteorological drought in a semiarid region using two long-term satellite-estimated rainfall datasets: A case study of the Piranhas River basin, northeastern Brazil. *Atmos. Res.* **2021**, *250*, 105380. [CrossRef]
12. Water, U.N. Water and climate change. In *The United Nations World Water Development Report*; Water UN: Geneva, Switzerland, 2020.
13. World Economic Forum. *This Is What Climate Change Costs Economies around the World*; World Economic Forum: Cologny, Switzerland, 2023; Available online: www.weforum.org (accessed on 25 March 2024).
14. World Meteorological Organization. *State of the Global Climate 2022*; World Meteorological Organization: Geneva, Switzerland, 2022; Available online: https://wmo.int/ (accessed on 25 March 2024).

15. Thomas, T.; Jaiswal, R.K.; Galkate, R.; Nayak, P.C.; Ghosh, N.C. Drought indicators-based integrated assessment of drought vulnerability: A case study of Bundelkhand droughts in central India. *Nat. Hazards* **2016**, *81*, 1627–1652. [CrossRef]
16. McKee, T.B.; Doesken, N.J.; Kleist, J. The relationship of drought frequency and duration to time scales. In Proceedings of the 8th Conference on Applied Climatology, Anaheim, CA, USA, 17–22 January 1993; Volume 17, pp. 179–183.
17. Vicente-Serrano, S.M.; Beguería, S.; López-Moreno, J.I. A multiscalar drought index sensitive to global warming: The standardized precipitation evapotranspiration index. *J. Clim.* **2010**, *23*, 1696–1718. [CrossRef]
18. Palmer, W.C. Meteorological drought. US. Weather Bureau Res. Paper, 1965. 45, 1–58.
19. Tsakiris, G.; Pangalou, D.; Vangelis, H. Regional drought assessment based on the Reconnaissance Drought Index (RDI). *Water Resour. Manag.* **2007**, *21*, 821–833. [CrossRef]
20. World Meteorological Organization. *Standardized Precipitation Index User Guide*; World Meteorological Organization: Geneva, Switzerland, 2012; p. 1090.
21. Mahmoudi, P.; Rigi, A.; Miri Kamak, M. A comparative study of precipitation-based drought indices with the aim of selecting the best index for drought monitoring in Iran. *Theor. Appl. Climatol.* **2019**, *137*, 3123–3138. [CrossRef]
22. Wable, P.S.; Jha, M.K.; Shekhar, A. Comparison of drought indices in a semi-arid river basin of India. *Water Resour. Manag.* **2019**, *33*, 75–102. [CrossRef]
23. Şen, Z. Fuzzy standardized precipitation index (FSPI) for drought early warning procedure. *Theor. Appl. Climatol.* **2023**, *155*, 1281–1287. [CrossRef]
24. Abu Arra, A.; Şişman, E. Innovative Drought Classification Matrix and Acceptable Time Period for Temporal Drought Evaluation. *Water Resour. Manag.* **2024**, *38*, 2811–2833. [CrossRef]
25. Şen, Z.; Şişman, E. Probabilistic standardization index adjustment for standardized precipitation index (SPI). *Theor. Appl. Climatol.* **2023**, *155*, 2747–2756. [CrossRef]
26. Elouissi, A.; Benzater, B.; Dabanli, I.; Habi, M.; Harizia, A.; Hamimed, A. Drought investigation and trend assessment in Macta watershed (Algeria) by SPI and ITA methodology. *Arab. J. Geosci.* **2021**, *14*, 1329. [CrossRef]
27. Gumus, V. Evaluating the effect of the SPI and SPEI methods on drought monitoring over Turkey. *J. Hydrol.* **2023**, *626*, 130386. [CrossRef]
28. Dabanlı, İ.; Mishra, A.K.; Şen, Z. Long-term spatio-temporal drought variability in Turkey. *J. Hydrol.* **2017**, *552*, 779–792. [CrossRef]
29. Burt, S.; Burt, T. *Durham Weather and Climate since 1841*; Oxford University Press: Oxford, UK, 2022; 580p.
30. Pettitt, A.N. A non-parametric approach to the change-point problem. *Appl. Stat.* **1979**, *28*, 126–135. [CrossRef]
31. Buishand, T.A. Some methods for testing the homogeneity of rainfall records. *J Hydrol.* **1982**, *58*, 11–27. [CrossRef]
32. Wijngaard, J.; Tank, A.; Können, G. Homogeneity of 20th Century European Daily Temperature and Precipitation Series. *Int. J. Climatol.* **2003**, *23*, 679–692. [CrossRef]
33. Wang, H.; Chen, Y.; Pan, Y.; Chen, Z.; Ren, Z. Assessment of candidate distributions for SPI/SPEI and sensitivity of drought to climatic variables in China. *Int. J. Climatol.* **2019**, *39*, 4392–4412. [CrossRef]
34. Stephens, M.A. Use of the Kolmogorov–Smirnov, Cramer–Von Mises and related statistics without extensive tables. *J. R. Stat. Soc. Ser. B: Stat. Methodol.* **1970**, *32*, 115–122. [CrossRef]
35. Yevjevich, V.M. Objective Approach to Definitions and Investigations of Continental Hydrologic Droughts. Ph.D. Thesis, Colorado State University, Collins, CO, USA, 1967.
36. Abu Arra, A.; Şişman, E. Investigation of the main difference between the run and SPI theories regarding drought characteristics. In Proceedings of the AHI EVRAN 3rd International Congress on Scientific Research, Baku, Azerbaijan, 3–4 May 2023; 2023; Volume 2, pp. 124–134.
37. Abu Arra, A.; Şişman, E. Characteristics of hydrological and meteorological drought based on intensity-duration-frequency (IDF) curves. *Water* **2023**, *15*, 3142. [CrossRef]
38. Peng, Y.; Peng, T.; Li, Y. Spatiotemporal Characteristics of Drought in Northwest China Based on SPEI Analysis. *Atmosphere* **2023**, *14*, 1188. [CrossRef]
39. Tigkas, D.; Vangelis, H.; Tsakiris, G. Implementing crop evapotranspiration in RDI for farm-level drought evaluation and adaptation under climate change conditions. *Water Resour. Manag.* **2020**, *34*, 4329–4343. [CrossRef]
40. Tsakiris, G. Drought risk assessment and management. *Water Resour. Manag.* **2017**, *31*, 3083–3095. [CrossRef]
41. Şişman, E. Self-similar characteristics of drought duration, total deficit, and intensity curves. *Arab. J. Geosci.* **2020**, *13*, 8. [CrossRef]
42. Tigkas, D.; Vangelis, H.; Tsakiris, G. Drought characterisation based on an agriculture-oriented standardised precipitation index. *Theor. Appl. Climatol.* **2019**, *135*, 1435–1447. [CrossRef]
43. Wang, W.; Wang, J.; Romanowicz, R. Uncertainty in SPI calculation and its impact on drought assessment in different climate regions over China. *J. Hydrometeorol.* **2021**, *22*, 1369–1383. [CrossRef]
44. Laimighofer, J.; Laaha, G. How standard are standardized drought indices? Uncertainty components for the SPI & SPEI case. *J. Hydrol.* **2022**, *613*, 128385.

*atmosphere*

MDPI

*Article*

# Assessment of Corn Grain Production Under Drought Conditions in Eastern Mexico Through the North American Drought Monitor

**Ofelia Andrea Valdés-Rodríguez [1], Fernando Salas-Martínez [2], Olivia Palacios-Wassenaar [3,*] and Aldo Marquez [4]**

1 El Colegio de Veracruz, Xalapa 91000, Veracruz, Mexico; andrea.valdes.colver@gmail.com
2 Área Académica de Química, Instituto de Ciencias Básicas e Ingeniería, Universidad Autónoma del Estado de Hidalgo, Carr. Pachuca-Tulancingo km. 4.5, Mineral de la Reforma, Pachuca 42184, Hidalgo, Mexico; fersamtz@gmail.com
3 Instituto de Ecología A. C., Xalapa 91070, Veracruz, Mexico
4 Área Académica de Computación y Electrónica, Instituto de Ciencias Básicas e Ingeniería, Universidad Autónoma del Estado de Hidalgo, Carr. Pachuca-Tulancingo km. 4.5, Mineral de la Reforma, Pachuca 42184, Hidalgo, Mexico; li.aldomg@gmail.com
* Correspondence: olivia.palacios@inecol.mx

**Abstract:** Over 80% of corn on Mexico's eastern side is sown under rainfed conditions. Therefore, drought represents a constant challenge for local producers. This study aims to determine the effects of drought on rainfed corn grain production on Mexico's eastern side by using the North American Drought Monitor as the primary tool. Drought levels at the municipal level provided by this monitor and corn production data (surface damage, yield, and volume) of the two productive seasons (spring–summer and autumn–winter) during 20 years were correlated at two significant levels (0.05 and 0.01). The significant values ($p < 0.05$) were used to obtain regression curves representing corn-drought behaviors. The National Disaster Statistics and climatological stations were considered, discarding other phenomena besides drought. Results indicate that, for the significant municipalities, the years with the highest drought levels (2005, 2011, and 2019) positively correlate with reduced corn grain yield, volume, and total harvest losses. The regression curves estimated a yield reduction of 78 kg·ha$^{-1}$ during the spring–summer season and 76 kg·ha$^{-1}$ during the autumn–winter season. We concluded that the Drought Monitor is valuable for determining relationships between rainfed corn grain productivity and drought, considering that no other climatological phenomena affect the region.

**Keywords:** *Zea mays*; productivity; drought; disasters; Veracruz; Tamaulipas

## 1. Introduction

### 1.1. The North American Drought Monitor

Drought is a low-evolution phenomenon that is challenging to predict due to its hazardous nature. It is characterized by abnormally dry conditions that cause a hydrological imbalance that can have destructive effects [1]. According to the impact drought causes, it can be classified as meteorological, agronomical, hydrological, and socioeconomic [2].

Since drought is considered an extreme event, the United States of America, Canada, and Mexico implemented a cooperative effort among their drought experts to continuously monitor drought across their regions. This cooperation gave birth to the North American Drought Monitor [3]. In Mexico, the National Meteorological Service named its side the


https://doi.org/10.3390/atmos16020193

Mexican Drought Monitor [4]. According to the North American Drought Monitor, drought levels are established by considering local monitors of air, fires, precipitations, soil moisture, streamflows, temperatures, vegetation indexes, and dams' water percentage [3]. Satellite images are also used to compose the tool and make the information available. This tool provides historical statistics and a map showing where drought occurs, its spatial extent, its severity, and the time scale of the associated impacts. It also includes information about the different drought levels in each municipality of the Mexican territory. According to this monitor, six possible levels represent drought conditions: (1) 0—No drought conditions were determined in the region. (2) D0—Abnormally dry: when it occurs at the beginning of a drought, short-term dryness may cause slow planting and growth of crops or pastures. When it occurs at the end of a drought period, it may cause lingering water deficits and pastures or crops not fully recovered (3) D1—Moderate drought: which may cause several crop and pasture damages, a high risk of fires, and low levels in rivers, streams, water reservoirs, and wells. (4) D2—Severe drought: associated with a high probability of crop damages, a high risk of fires, water shortage, and limits in water consumption. (5) D3—Extreme drought: indicates high crop or pasture losses, extreme risk of forest fires, and a general restriction in water consumption. (6) D4—Exceptional drought: indicates widespread crop or pasture losses and shortages of water reservoirs, streams, and wells, creating water emergencies [3]. With these levels, the Drought Monitor provides 15-day reports for each municipality of the Mexican territory, which can be used by decision-makers, producers, and ordinary citizens interested in the evolution of drought for research purposes [4]. Considering these definitions, it can be assumed that meteorological, agricultural, and hydrological droughts are the primary drought types for the Drought Monitor [5].

### *1.2. Historical Drought Levels in Mexico and Their Relationship with Agricultural Data*

Mexico is a country where drought usually affects more than 50% of its territory [6]. Nowadays, this situation worsens if we consider that drought periods are increasing in more extensive parts of the country. For example, according to the last report of the Weather Meteorological Organization in Latin America and El Caribe, drought and high temperatures in 2022 have increased in average values [7].

The historical records of the colonial period and the modern Mexican Drought Monitor [4] indicate that Mexico has experienced periodical droughts [8]. Nevertheless, the Drought Monitor shows that there have been more increments in their intensity since 2011, when more than 85% of the territory recorded D1 or above levels of drought. Regarding agricultural data, old records indicate that severe drought has caused the loss of multiple crops in the southeastern side of Mexico (mainly in the Yucatan peninsula), forcing people to displace and leave their home cities since pre-colonial and colonial times [9]. More recently, three significant dry periods were registered in the Mexican territory for the last century: 1948–1954, 1970–1978, and 1993–1998. During the first period, the northeastern side of Mexico (above the parallel 22°) reported 250,000 ha of cotton lost by the drought. The eastern side of Mexico was entirely affected during the second period, causing the total loss of corn and cotton sown. For the last period, all the territory was affected by drought. During this period, corn and wheat were reported to have the highest damaged surfaces, with more than 300,000 ha lost. Although this information does not mention if the affected surfaces were rainfed or irrigated lands, these reports specify that corn cultivation generally recorded the highest surface damage during drought in eastern Mexico [2].

### *1.3. Corn Production in Mexico*

Corn (*Zea mays* L.) is the most cultivated crop in Mexico, sown in every state of the Mexican Republic [10,11]. It can be planted under different agroclimatic conditions, from sea level up to 3000 m above sea level and in wet to dry climates [12]. Therefore, corn cultivation is subject to many hydrometeorological phenomena, like excessive rains, strong winds, cold temperatures, and droughts. In the country, corn has different applications: corn grain, which is sold as cob grain; fodder corn, which is sold as pasture for animal feed; popcorn corn, a unique variety for human consumption; and corn seed, sold for sowing purposes [11]. From these varieties, corn grain is the most cultivated variety in the country, with more than 27.50 tons harvested in 2022 [13]. It is also used for animal and human consumption, and the yellow and white varieties are the most common sown in the country.

### 1.4. Corn Grain on the Eastern Side of the Mexican Territory

On the eastern side of Mexico, along the Gulf of Mexico, most of the territory includes two states: Veracruz in the south and Tamaulipas in the north. This area combines a mix of tropical, subtropical, semidry, and dry climates, where agricultural activities are an essential part of living for many local producers. Despite climate conditions, seasonal rainfed crops, which are sown during specific periods of the year and depending on local climatic conditions, are sown yearly, regardless of possible drought conditions. In Mexico, according to the National Service for Agrifood and Fisheries Information (SIAP), there are two seasons for sowing and harvesting: spring–summer and autumn–winter. The spring–summer period lasts from April to September, while the autumn–winter period lasts from October to March. In this context, corn is sown in most Mexican territories in both seasons [11].

In both states (Veracruz and Tamaulipas), corn grain is sown under seasonal conditions in 247 municipalities out of their 255 locations. This situation represents a yearly average of 444,861 hectares of land under rainfed corn grain cultivation [11]. Previous research on the performance of different crops correlated with climatic phenomena in part of the eastern side of Mexico indicates that tropical cyclones and floods have the highest number of significant correlations with damaged cornland. However, this study also shows that drought is in third place, causing total rainfed corn grain losses of 63,705 ha during the most recent worst drought periods registered on this site [14]. In other studies on Mexico's western side, corn has also been affected by prolonged droughts, with a 43% reduction in crop grain productivity during the dry seasons [15]. One local study in the east determined a positive correlation between drought and corn productivity; however, the area under study was only a tiny fraction of the eastern central side of Mexico (8605 $km^2$) [16]. Besides this, according to future climate change scenarios [17], there will be an increase in drought periods and temperatures in the Mexican territory.

Since corn is an essential food of the Mexican diet, its production and its relationship with climate data require an analysis where it is necessary to access local climatologic records and corn productivity per site. In this regard, the National Meteorological Service of Mexico (SMN) is the governmental entity responsible for maintaining climatic records from the Mexican territory [18]. The SMN contains historical data since the beginning of each local station installation. However, the eastern side of Mexico has poor coverage, with only approximately 31% of the information accurately updated [19]. This situation is critical to determining climatic conditions because this region has a complex geography. First, the whole area has more than ten longitudinal degrees of length, and second, mountain chains cross from south to north along the territory, with some sites at sea level and, in contrast, others up to 4000 m above sea level (masl) in less than 200 km of distance from east to west [20,21].

Due to this lack of information, one study aimed to predict agricultural drought behavior in the short term for the central part of the Gulf of Mexico using a Long Short-

Term Memory Network and Landsat 8–9 multispectral images. It required the analysis of 183 high-resolution images for an area of only 5177.9 $km^2$ for 10 years. However, the prediction level was low when high and low droughts occurred in regions a few kilometers away [22]. These prediction difficulties can be explained by another study suggesting that 30 or more samples should be used to feed a model [23]. In this case, this model additionally would need to consider different seasonal and agricultural periods. For example, when the land is uncovered because of the tillage, when crops cover it completely, and the natural seasons of the year. These different land covers increase the difficulties in detecting drought levels in territories dedicated to agronomical activities.

Therefore, considering the extensive territory of the eastern Mexican side and the difficulties of analyzing drought indexes in municipalities without climatological information, the drought levels provided by the Drought Monitor are the only ones available for all the municipalities cultivating corn grain. The Drought Monitor is also considered the most holistic measure of drought [5]. Besides this situation, corn grain productivity and its relationship with drought levels, like the ones reported by the North American Drought Monitor, have not been studied yet in a land of 152,073 $km^2$, which is the one that comprises Veracruz and Tamaulipas. For example, the North American Drought Monitor periodically reports drought conditions in cornfields in the United States of America [24], which is absent on the Mexican side.

In addition, there is no information regarding drought levels occurring during governmental Declarations of Disaster (DD) by drought. A DD by drought is issued when drought levels cause crop or cattle losses or when the habitant's capacity to obtain water for their primary and domestic uses is compromised [25]. Therefore, learning the link between drought levels and corn grain reduction and losses will allow producers to evaluate historical corn development under arid conditions, especially considering climate change.

This research aims to evaluate corn grain productivity under different drought levels provided by the Mexican Drought Monitor on the eastern side of Mexico (Veracruz and Tamaulipas states).

## 2. Materials and Methods

### *2.1. Research Site and Analysis Period*

The total territory is situated at 27°40′45″ north latitude, 100°08′42″ west longitude, 17°10′00″ south latitude, and 93°36′29″ east longitude [20,21]. On the southern side of the territory (below the parallel 23°), the more considerable extension has a warm subhumid climate, with summer rains, in 54.13% of the territory, while on the northern side of the territory, a semiwarm subhumid climate, with summer rains, occupies 29.38% of the territory, followed by a semidry warm climate, with summer rains, in 23.29% of the territory. Other climate types, like temperate humid, warm dry, temperate dry, and semicold subhumid, comprise 3.2% of the territory.

Since the agrifood system databases and the Drought Monitor used for this research contain historical information at the municipal level starting in 2003, this research considers only data from 2003 to 2022 (20 years), the available information period for both databases.

### *2.2. Extreme Climatic Conditions Causing Disaster Declarations*

Besides the Drought Monitor, extreme climatic conditions were considered in this study because they may affect crop fields during their occurrence, and, in these cases, the Mexican government issues a Disaster Declaration at a municipal level. The National Center for Disaster Prevention (CENAPRED) is the official agency that maintains a database with DD containing the dates and climatic phenomena causing the catastrophe [25]. Therefore, data can be obtained from CENAPRED and compared with corn production to determine other phenomena besides drought affecting the crop. Table 1 defines the most common climatic phenomena registered by CENAPRED and the classification this research considers.

**Table 1.** Classification of climatic phenomena causing disaster declarations in Mexico.

| Identification | Climatic Conditions |
|---|---|
| Tropical cyclones | They can be tropical storms, tropical cyclones, and hurricanes. |
| Rains | Heavy or excessive precipitation that was not expected and caused severe losses. |
| Floods | Floods occur when a river, lake, or lagoon overflows. They can also happen when a specific place is flooded, causing losses to properties, people, cattle, or crops. |
| Low temperatures, snow, frost, or hail | They can include low temperatures, snow, frost, or hail, causing damage to people or crops. |
| Winds | These are strong winds destroying properties or crops. |
| High temperatures | They are extreme temperatures causing health problems for humans and animals. |
| Droughts | Droughts are considered when rains do not occur during their typical period. This phenomenon causes low water reservoirs, lack of water, and cattle and crop losses. |

Source: National Center for Disaster Prevention [25].

### 2.3. *Meteorological Information*

The meteorological information considered temperature and precipitation obtained from the national meteorological stations in the area [26]. This information was used to explain the historical climatological behavior of the sites where meteorological information is available. Of the 283 available meteorological stations operating in the area, only 38 contained the minimum 80% of data recommended by the World Meteorological Organization [27] for the 42-year base period (1980–2022). This extended period was considered because the base period reported by the Mexican Meteorological Service started in 1980 and concluded in 2010. However, significant drought levels have been recorded in recent years (2011, 2013, and 2019) [4], which were considered in this study.

Figure 1 shows the location of the available meteorological stations in the study region and its most representative climate types. The information on each station is shown in Table A1.

### 2.4. *Corn Productivity Data*

Data for this research considered the 247 municipalities in Veracruz and Tamaulipas with historical records of corn grain production. Corn grain data were obtained from the Mexican National Service for Agrifood and Fisheries Information (SIAP) [11]. This service has provided municipal data since 2003. Data available consider the surface of land sown (ha), the surface damaged (although without information about the cause), the surface of land harvested (ha), volume harvested (tons), yield in tons by a hectare ($tons{\cdot}ha^{-1}$), and the selling price by ton (Mexican pesos $tons^{-1}$). This data is provided for each season (spring–summer and autumn–winter).

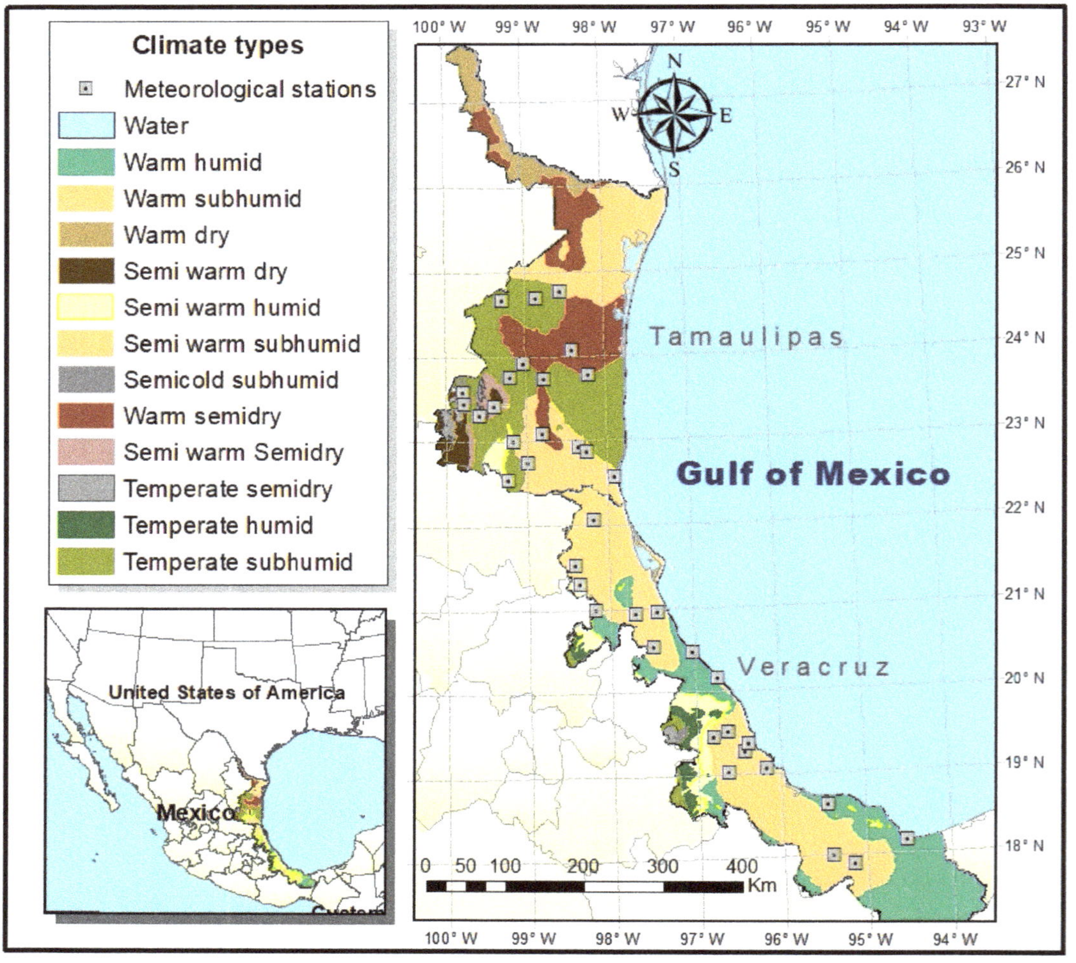


**Figure 1.** Study region, climate types, and available meteorological stations. Source: Climate and cartographic data were obtained from the latest update of the Mexican Institute of Geography and Statistical Information [28]; meteorological station locations were obtained from the National Meteorological Service [26].

### 2.5. Drought Levels and Drought Intensity Estimations

Historical drought levels at the municipal level were obtained from the National Drought Monitor of Mexico, which started in January 2003. The drought levels provided by the Mexican Drought Monitor were converted to numerical values as follows: 0 = no drought condition, 1 = D0, 2 = D1, 3 = D2, 4 = D3, and 5 = D4, to quantify drought intensity.

Since corn grain production data considers the whole six-month period (spring–summer and autumn–winter seasons), equivalent drought levels were estimated for the same period according to Equation (1).

$$It_{season} = (\Sigma D0 \times 1 + \Sigma D1 \times 2 + \Sigma D2 \times 3 + \Sigma D3 \times 4 + \Sigma D4 \times 5)/m \quad (1)$$

where $It_{season}$ is the ponderate seasonal drought level of the municipality $n$.

$\Sigma D0$, $\Sigma D1$, $\Sigma D2$, $\Sigma D3$, and $\Sigma D4$ represent the sum of drought levels reported by the Drought Monitor at $n$ municipality, and $m$ is the number of reports during the seasonal period.

Additionally, the mean drought level by season per site period (the average drought level over the 20 years) was estimated to consider positive and negative oscillations.

Equation (2), which estimates the period's summary drought intensity by each season ($Itot_{season}$), was used to obtain the municipalities with the highest drought records from 2003 to 2022.

$$Itot_{season} = \Sigma D0 \times 1 + \Sigma D1 \times 2 + \Sigma D2 \times 3 + \Sigma D3 \times 4 + \Sigma D4 \times 5 \quad (2)$$

The level of drought occurring during a Declaration of Disaster by drought was evaluated considering the maximum and average drought level recorded by the Drought Monitor during the period the statement was issued in the affected municipality.

### 2.6. Corn Yield and Volume Variation by Year and Season

Corn grain data per site summarized the number of hectares sown, harvested, or damaged by all the communities in the municipality. Corn grain yield was estimated as the average of all the municipalities (tons harvested over sown surface). Each period's average values (surface sown, surface harvested, surface damaged, yield, and volume) per site were estimated to determine the total performance of the 2003–2022 period.

Linear regressions were estimated to determine the performance of yield and volume values during the analysis period. The regressions indicated that yield and volume values increased periodically during the study period. We assumed this increase could originate from new technologies or corn varieties. Therefore, for this study, the yield and volume per year and season were estimated using only the previous year of data, as shown in Equation (3).

$$yValue = current\ year\ value - previous\ year's\ value \quad (3)$$

where *yValue* can be the yield or volume decrement or increment recorded during the current year, the *current year value* can be the yield or volume recorded during the current year, and the *previous year's value* means the yield or volume of the prior year.

This way, we obtained the data series (yield and volume variation by year and season) to compare drought intensity by year and season.

### 2.7. Damaged Corn Surfaces and Drought Levels

Only years with drought declarations and the municipalities affected were considered in determining the correlation between damaged corn surfaces and drought levels because the study region registers many climatological phenomena other than drought. Drought levels and their intensities were considered using Equations (1) and (2) to correlate with damaged corn surfaces.

### 2.8. Summary of Sources of Information and Data Resolution

Table 2 summarizes the databases, period of study, and the spatial resolution of data used in this study for better clarification. The land use and climate types of cartographic information were considered to complement the discussion over the impacts of drought on corn grain productivity in relationship with their climate types and areas of agricultural land use.

**Table 2.** Database sources for the information consulted during this research.

| Data | Period | Spatial Resolution | Source |
|---|---|---|---|
| Climatological data | 1980–2022 | By local meteorological station | Mexican National Meteorological Service [26] |
| Land use | 2020 | 1:250,000 | Mexican Institute of Geography and Statistical Information [28] |
| Climate type | 2001 | 1:250,000 | Mexican Institute of Geography and Statistical Information [28] |
| Drought levels | 2003–2022 | By municipality | Mexican Drought Monitor is provided by the National Meteorological Service [4] |
| Disaster Declarations | 2003–2022 | By municipality | Mexican National Center of Disaster Prevention [25] |
| Corn grain productivity | 2003–2022 | By Locality | Mexican National Service for Agrifood and Fisheries Information [11] |

### *2.9. Data Analysis*

The Pearson correlation was used to determine the correlation coefficients ($r$) between drought levels and the number of declarations by drought and between crop data and drought levels. The statistical tests were performed using the software SigmaPlot 11.0.

For both seasons (spring–summer and autumn–winter), Disaster Declarations by drought, corn production (yield and volume), and its relationships with drought levels by municipality and year were correlated with a 0.05 and 0.01 significance. The Argis Desktop 10.8 software was used to show results with cartographic information. Since the damaged surfaces registered by the SIAP do not provide the reasons for the damage, only the municipalities with Disaster Declarations by drought provided by CENAPRED were considered to exclude other phenomena besides drought, as explained in Table 1.

Only data with significant correlations ($p < 0.05$) were selected to estimate the regression equations with the best fit ($r^2$) to determine drought intensity in relation to corn grain yield and volume production and drought intensity versus damaged surface by using the SigmaPlot 11.0 program. Figure A1 shows the flow diagram of the analytical process.

## 3. Results

### *3.1. Drought Levels of the Study Site*

Figure 2a shows the study region's average drought levels per season. During the spring–summer season, the maximum average drought level was D1. During the autumn–winter season, two municipalities on the northeastern territory registered average D2 levels. However, only 9.0% of municipalities did not report drought during spring–summer, while 40.8% did not report drought during the autumn–winter. Figure 2b shows the summary of drought intensities of the municipalities during the period 2003–2022. Above parallel 25°, the northern side of the region has the highest drought levels during spring–summer, followed by the southern region, below parallel 18°. The highest sum of drought levels (351) on the northern side, above parallel 26°, indicates that 58% of the records were D3 and D4 drought levels.

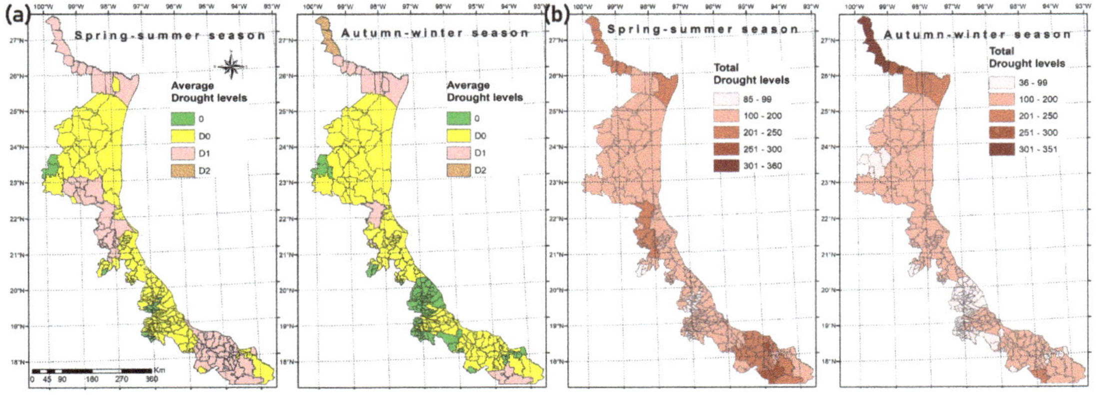


**Figure 2.** Drought intensities in the study region. (**a**) Estimated average drought levels and their intensity in the study region during 2003–2022. (**b**) Sums of drought levels ($Itot_{season}$) recorded by municipality during 2003–2022. 0: no drought, D0: abnormally dry, D1: moderate drought, D2: severe drought. Sources: geographic information was obtained from the Mexican Institute of Geography and Statistical Information [29]. Drought data was obtained from the Mexican Drought Monitor [4].

Figure 3 shows the historical drought levels recorded in the study site. It can be seen that 2005, 2011, 2019, and 2020 were the years with the highest drought levels. The spring–summer season generally accounts for the highest drought levels (Figure 3a). For example, during spring–summer 2011, 47% of the municipalities recorded D3 drought levels, and 2% (six municipalities) recorded D4 levels, while during the autumn–winter season, only 4% of the sites recorded D3 levels, and D4 levels were not recorded (Figure 3b). During spring–summer 2019, 37% of the sites recorded D3 levels and 4% D4; during autumn–winter, only 6% recorded D3 levels, and D4 levels were not registered. By 2020, during the spring–summer season, from June to September, the number of municipalities without drought decreased from 93% to 35%, descending to 25% in April 2021 and 5% in May 2022.

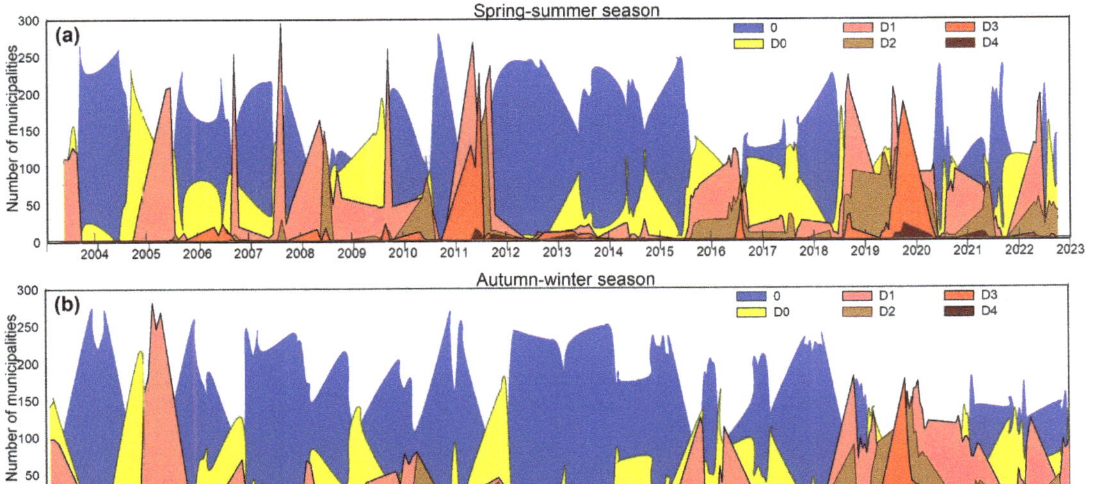


**Figure 3.** Historical drought records in the study region during (**a**) spring–summer and (**b**) autumn–winter. 0: no drought, D0: abnormally dry, D1: moderate drought, D2: severe drought, D3: extreme drought, D4: exceptional drought. Source: drought data was obtained from the Mexican Drought Monitor [4].

### 3.2. Disaster Declarations and Drought Levels

Although there have been only five years with records of Disaster Declarations caused by drought, the levels of drought registered in the affected municipalities differed widely. Figure 4 shows the levels of drought registered during the 200 Disaster Declarations for the autumn–winter season and 132 recorded for the spring–summer season during 2003–2002. The lowest average level of drought recorded during Drought Declarations was D0, with only three municipalities indicating this level. In contrast, the highest average drought level recorded was D4, with nine sites recording it. The correlation coefficients between drought levels and Disaster Declarations by drought were significant at $p < 0.01$ for the autumn–winter season ($r = 0.303$) and the spring–summer season ($r = 0.340$).

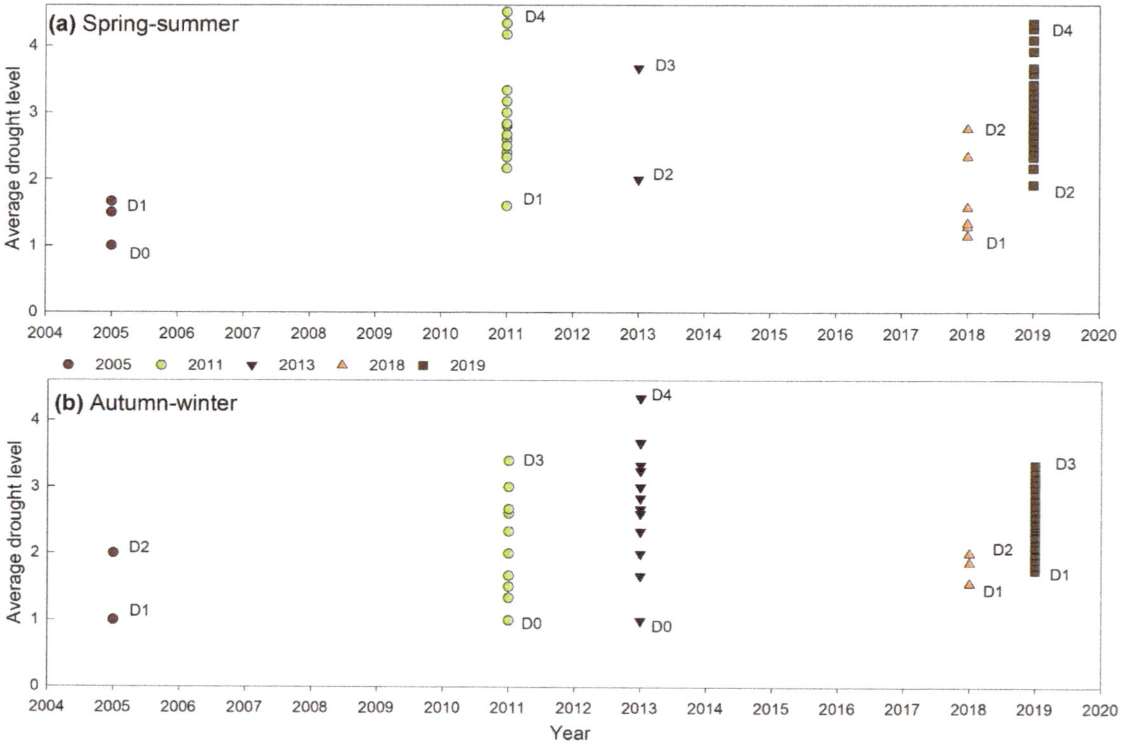


**Figure 4.** Drought levels and Disaster Declarations by drought registered in the affected municipalities of the study region from 2003 to 2022. (**a**) spring–summer season; (**b**) autumn–winter season. D0: abnormally dry, D1: moderate drought, D2: severe drought, D3: extreme drought, D4: exceptional drought. Sources: drought data were obtained from the Mexican Drought Monitor [4], and Disaster Declarations by drought were obtained from CENAPRED [25].

Figure 5 shows the number of Disaster Declarations by drought in each municipality and its corresponding average drought level recorded during the declarations. The autumn–winter season registered 60% of the total number of Disaster Declarations of the two seasons, with 100% of the municipalities having at least one Disaster Declaration due to drought above the parallel 22.5° and two sites having up to three Disaster Declarations (Figure 5a). Nevertheless, the drought levels during the spring–summer season were higher than during the autumn–winter season, with average levels up to D2, D3, and D4 above parallel 25°. In the South, below parallel 18°, three sites reached D2 levels (Figure 5b). The highest number of Disaster Declarations by drought in two municipalities (Tantoyuca and Camarón de Tejeda (Figure 5a) was three during the autumn–winter season and two during the spring–summer season.

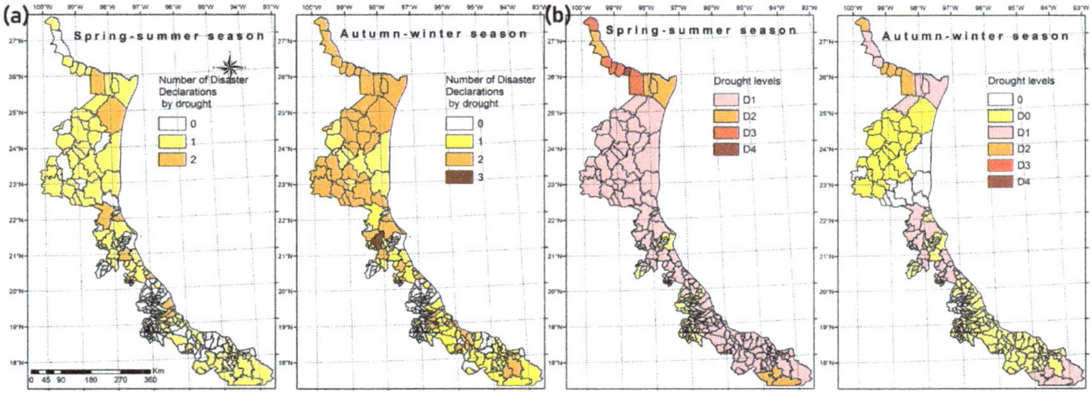


**Figure 5.** (**a**) Number of Disaster Declarations by drought registered during 2003–2022 and (**b**) the corresponding average drought levels recorded during the Disaster Declarations of 2005, 2011, 2013, 2018, and 2019. Geographic data was obtained from the Mexican Institute of Geography and Statistical Information [28].

Figure 6 shows the monthly temperature behavior during four years with Disaster Declarations by drought: 2005, 2011, 2013, and 2019, in relation to the long-term climatology (1980–2022). During 2005 (Figure 6a). Positive anomalies occurred during the spring–summer (June—September) and January (autumn–winter season). These positive anomalies oscillated between 2% and 8% of the total range and represented temperature increments between 0.2 °C and 1.4 °C, with average increments of 0.3 °C for the spring–summer period. These increments were located above the 75th percentile.

For 2011 (Figure 6b), there were seven months with positive anomalies oscillating between 3% and 9% of the total range, implicating temperature increments between 0.3 °C and 2.0 °C. These increments were recorded during two months of the autumn–winter period (January and March) and five months of the spring–summer period, with average increments of 0.9 °C. These increments were located above the 75th percentile.

For 2013 (Figure 6c), February registered the highest temperature anomaly, 1.5 °C above the average, followed by October at 4% above the average and June and July at less than 3%. The other months did not register positive temperature anomalies.

Only March and April registered temperatures below the long-term average during 2019 (Figure 6d). The other months had positive anomalies from 1 to 6% above the media, with an average temperature increment of 0.7 °C for this year. During the spring–summer season, the temperature was 1.6 °C higher than the mean of the long-term period, corresponding to increments above the 75th percentile of the total period.

Figure 7 shows the monthly precipitation behavior during four years with Disaster Declarations by drought: 2005, 2011, 2013, and 2019 in relation to the long-term climatology (1980–2022). During 2005 (Figure 7a), the autumn–winter season (November, December, and January) and the spring–summer season (April and September) recorded negative anomalies. These anomalies oscillate between −10 and −80%, representing −5.7 and −54.6 mm of accumulative monthly rain. Nevertheless, October and February registered higher precipitations, up to 110% above the normal, representing 127.8 mm more than the average.

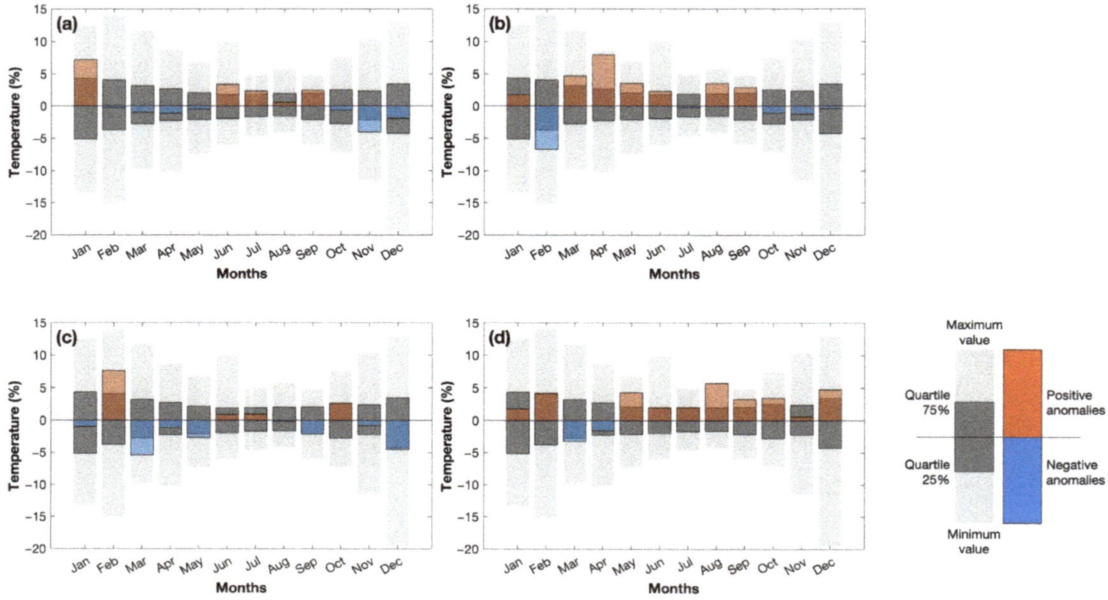


**Figure 6.** Monthly temperature behavior registered during the years with Disaster Declarations by drought. (**a**) 2005, (**b**) 2011, (**c**) 2013, and (**d**) 2019.

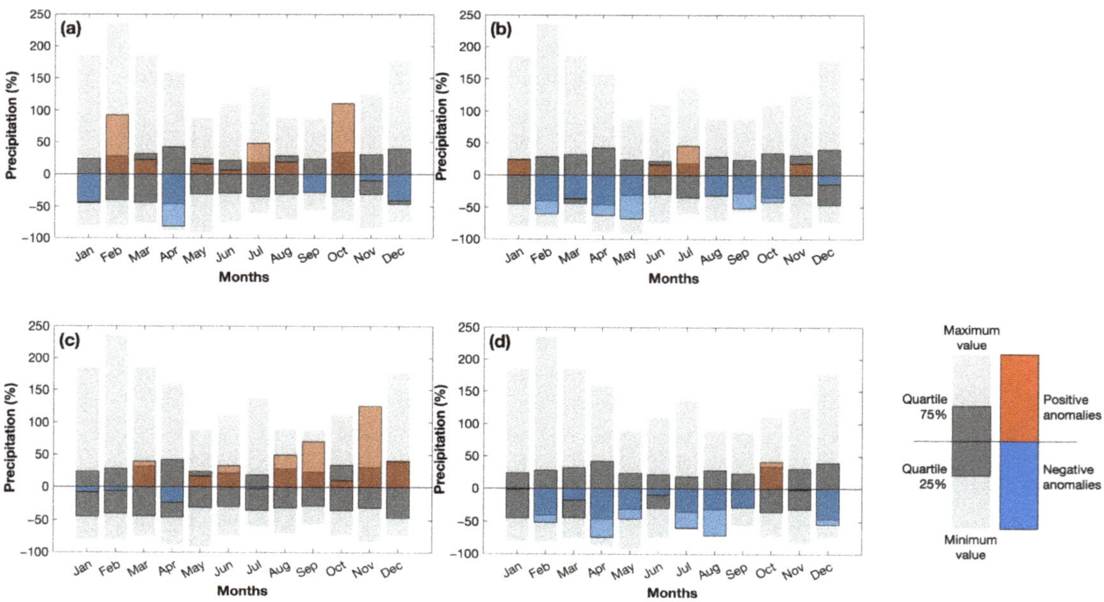


**Figure 7.** Monthly precipitation behavior registered during the years with Disaster Declarations by drought. (**a**) 2005, (**b**) 2011, (**c**) 2013, and (**d**) 2019.

In 2011 (Figure 7b), negative anomalies were recorded in eight months, ranging from −20 to −70%, representing −6 to −100 mm of less accumulative rain. Positive anomalies were recorded in January, June, July, and November, ranking from +20 to +45% of the normal, representing up to 69 mm more accumulative rain during July, with the highest precipitation records. Nevertheless, the negative anomalies were below the 75th percentile in more than 50% of the year. The average reduced precipitation recorded for the spring–summer season was −126.2 mm, while the autumn–winter season registered −58.6 mm less than the normal. The total reduction in precipitation during 2011 was 18% lower than the average of the long-term period.

In 2013 (Figure 7c), eight months had positive anomalies, with precipitations from 10 to 120% higher than average, and November being the month with the highest precipitations. These positive anomalies represent up to 254 mm more accumulative rain for the spring–summer season and 104 mm for the autumn–winter season. The negative anomalies were recorded only during January, February, and April, with average reductions of −10 mm, representing between −5 and −25% less rain than the long-term average.

In 2019 (Figure 7d), 11 months registered negative anomalies, and only October registered positive ones. These values oscillated between −2 and −75%, with precipitations reduced between −1 mm and −102 mm. During this year, six months had accumulative rainfall below the 25th percentile. Therefore, 2019 registered a total reduction of 30% below the long-term average.

### 3.3. Corn Grain Production

Figure 8a,b shows each municipality's average rainfed corn grain sown during the study period (2003–2022). The spring–summer season has the highest proportion of land, with 68% of the total surface of the year cultivated during this season. Below latitude 20°, the southern region has five municipalities cultivating more than 10,000 ha, while the northern has only one municipality during the spring–summer season. During the autumn–winter season, only one municipality cultivated more than 10,000 ha in the region. Figure 8c,d shows the volume obtained in the study region. Below parallel 21°, three municipalities produced more than 20,000 tons; above the parallel 22°, only three sites harvested more than 10,000 tons but less than 15,000 tons.

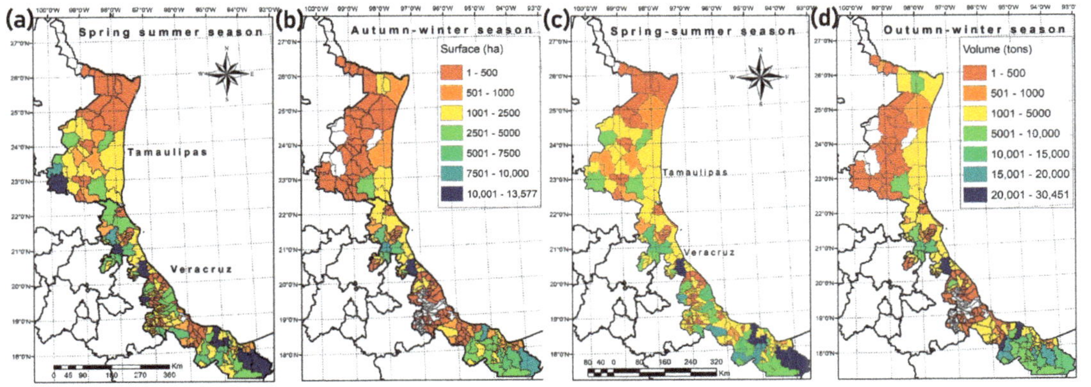


**Figure 8.** Average corn sown and volume obtained in the study region during 2003–2022. (**a**) Surface sowed during the spring—summer. (**b**) Surface sowed during the autumn—winter. (**c**) Volume harvested during spring—summer. (**d**) Volume harvested during the autumn—winter. Sources: Geographic data was obtained from the Mexican Institute of Geography and Statistical Information [28]. Production data obtained from the Mexican National Service for Agrifood and Fisheries Information [11].

### 3.4. Drought Levels and Variations in Corn Grain Productivity

There were negative significant correlations between drought levels and yield values in 11 of 143 municipalities for the autumn–winter season ($p < 0.01$). In comparison, there were ten negative significant correlations of 231 municipalities between drought levels and yield for the spring–summer season (Figure 9a). Six affected municipalities were above the 22.5° parallel during spring–summer and four below the parallel 20°. During the autumn–winter season, only one site with negative correlations was below the parallel 21°. There were significant negative correlations between drought levels and volume values in four municipalities during the spring–summer season and 10 for the autumn–winter season (Figure 9b). During the spring–summer season, two sites were below parallel 19°; during the autumn–winter season, three sites with negative significant correlations were located below parallel 21°.

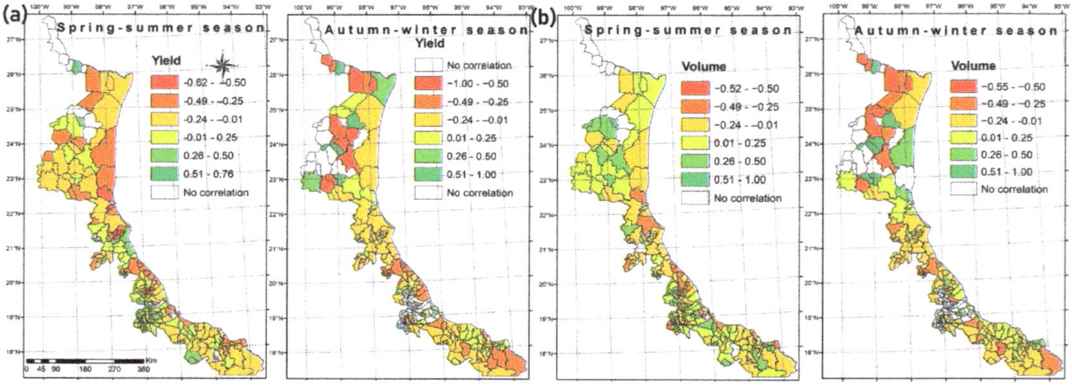


**Figure 9.** Coefficient correlations between corn grain yield variation, volume variation, and drought levels in the study region during 2003–2022. The blank areas indicate no correlation. (**a**) Yield variation; (**b**) volume variation. Sources: Geographic data was obtained from the Mexican Institute of Geography and Statistical Information [28]. Production data obtained from the Mexican National Service for Agrifood and Fisheries Information [11].

Correlations between corn grain yield variations and drought intensity registered nine statistically significant years for the spring–summer and eight for the autumn–winter season. Four years were statistically significant for the corn grain volume variation during the spring–summer season and five years for the autumn–winter season (Table 3). For the spring–summer season, in two years with positive correlations (2005 and 2014), the drought levels were below the total median of the 2003–2022 period. However, in 2016, 2020, 2021, and 2022, the yields experienced higher increments, independent of the drought levels. For the autumn–winter season, the years with negative correlations in corn grain yield corresponded with increments in the drought levels above the long-term average and vice versa. Similarly, the years with drought levels above the average corresponded with significantly different decrements in the production volume and vice versa (Figure 10).

**Table 3.** Correlations between corn grain production variations and drought levels during 2003–2022.

| Year | Spring-Summer Season | | | | Autumn-Winter Season | | | |
|---|---|---|---|---|---|---|---|---|
| | Yield | | Volume | | Yield | | Volume | |
| | r | p | r | p | r | p | r | p |
| 2004 | 0.11 | 0.08 | 0.13 | 0.04 * | −0.41 | 0.00 ** | −0.30 | 0.00 * |
| 2005 | 0.22 | 0.00 ** | −0.13 | 0.03 * | 0.19 | 0.03 | 0.11 | 0.07 |
| 2006 | −0.36 | 0.00 ** | −0.16 | 0.01 * | 0.19 | 0.02 * | −0.09 | 0.17 |

**Table 3.** *Cont.*

| Year | Spring-Summer Season | | | | Autumn-Winter Season | | | |
|---|---|---|---|---|---|---|---|---|
| | Yield | | Volume | | Yield | | Volume | |
| | r | *p* | r | *p* | r | *p* | r | *p* |
| 2007 | −0.24 | 0.00 ** | −0.07 | 0.30 | −0.24 | 0.00 ** | −0.09 | 0.13 |
| 2008 | 0.14 | 0.03 | 0.02 | 0.81 | 0.27 | 0.00 ** | 0.03 | 0.67 |
| 2009 | 0.15 | 0.03 | 0.12 | 0.05 | 0.09 | 0.32 | 0.13 | 0.03 * |
| 2010 | −0.28 | 0.00 ** | −0.42 | 0.00 * | −0.02 | 0.81 | −0.09 | 0.15 |
| 2011 | −0.15 | 0.02 * | −0.24 | 0.00 * | −0.24 | 0.01 ** | −0.07 | 0.26 |
| 2012 | 0.03 | 0.60 | −0.13 | 0.05 | −0.10 | 0.27 | −0.07 | 0.29 |
| 2013 | 0.04 | 0.50 | 0.04 | 0.50 | 0.08 | 0.33 | 0.04 | 0.49 |
| 2014 | 0.15 | 0.02 * | 0.13 | 0.04 * | −0.04 | 0.66 | −0.04 | 0.49 |
| 2015 | −0.02 | 0.75 | −0.06 | 0.34 | −0.21 | 0.01 ** | −0.06 | 0.37 |
| 2016 | 0.23 | 0.00 ** | 0.12 | 0.06 | 0.03 | 0.67 | 0.01 | 0.84 |
| 2017 | −0.02 | 0.76 | −0.01 | 0.89 | −0.19 | 0.02 * | −0.17 | 0.01 * |
| 2018 | −0.07 | 0.28 | −0.16 | 0.01 * | −0.07 | 0.40 | −0.14 | 0.02 * |
| 2019 | −0.24 | 0.00 ** | −0.04 | 0.52 | −0.01 | 0.88 | 0.05 | 0.45 |
| 2020 | 0.06 | 0.34 | 0.10 | 0.13 | 0.00 | 0.99 | −0.13 | 0.04 * |
| 2021 | 0.09 | 0.16 | 0.11 | 0.08 | −0.18 | 0.03 * | −0.14 | 0.02 * |
| 2022 | 0.29 | 0.00 ** | −0.11 | 0.07 | 0.24 | 0.00 ** | 0.26 | 0.00 ** |

* Statistically significant ($p < 0.05$), ** statistically significant ($p < 0.01$).

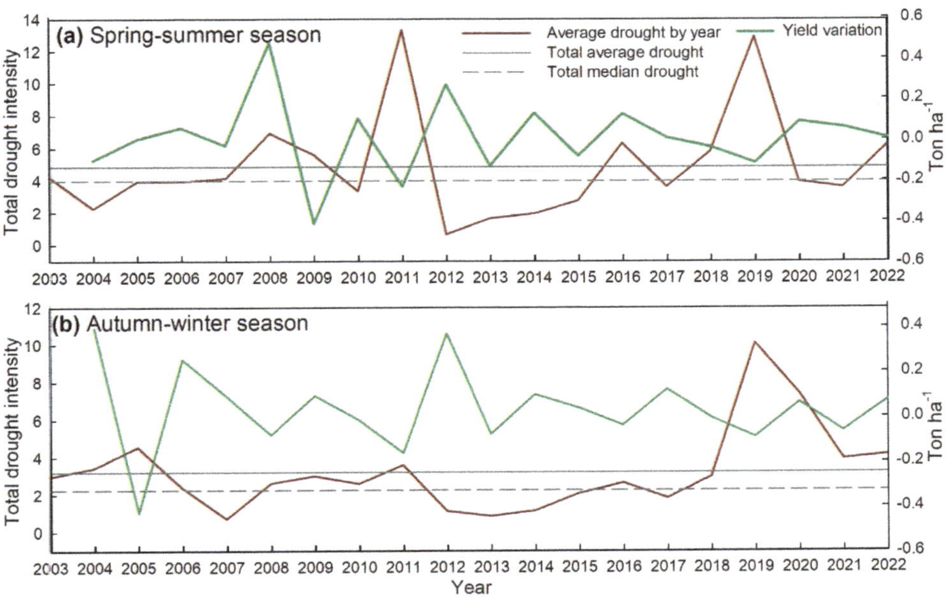


**Figure 10.** Corn grain yield variation and drought levels registered in the study zone during 2003–2023. (**a**) For the spring–summer season; (**b**) for the autumn–winter season.

### 3.5. Corn Grain Productivity Versus Drought Intensity

Figure 10 shows the performance of yield and volume production during the spring–summer and autumn–winter seasons versus the total drought intensity levels. In all cases, yield and volume decrease when drought intensity increases. The higher determination coefficient was obtained with linear regression, and the highest correlation was observed for the yield during the spring–summer season (Figure 11a). With the highest sum of drought intensities (14.33) during the spring–summer season, the yield can decrease to $-1.77$ tons·ha$^{-1}$. This total drought intensity level represents D2 and D3 levels recorded during at least 50% of the period of six months. During the autumn–winter season (Figure 11b), the decrements are slightly lower, with up to $-1.38$ tons·ha$^{-1}$ for average drought levels of D2. The volume production indicates that the autumn–winter season is more affected by drought intensity (Figure 11d), with 22% higher decrements during autumn–winter than during spring–summer (Figure 11c) and 11% lower intensity levels during the autumn–winter season.

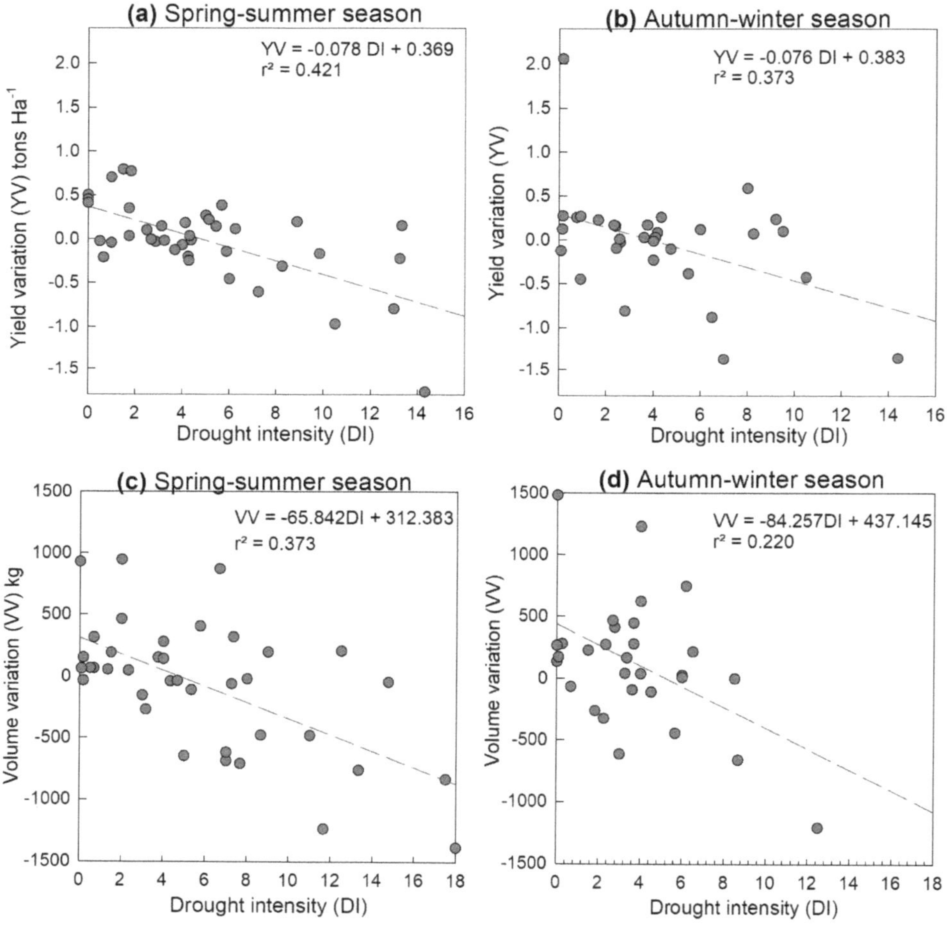


**Figure 11.** Drought levels versus yield and volume production of corn during 2003–2022 on the eastern side of Mexico. $r^2$ represents the determination coefficient of the regression line. (**a**) Spring–summer yield variation; (**b**) autumn–winter yield variation; (**c**) spring–summer volume variation; (**d**) autumn–winter volume variation. The green dots correspond to the intersection between the sum of the drought levels and the corresponding production yield or volume. The dashed gray lines are the results of the regression equations.

### 3.6. Damaged Corn Surfaces and Drought Intensity

Figure 12 shows the damaged surfaces of the corn grain and the land uses from 2003 to 2022. During the spring–summer season, the years with Disaster Declarations by drought reported 86,449.88 ha lost; during the autumn–winter season, only 18,412.87 ha were lost. This data corresponds to 4.70 times more damaged surfaces in spring–summer than in autumn–winter. The two municipalities with the highest damaged surfaces during the spring–summer season (more than 10,000 ha) were Chicontepec, Veracruz, and Tula, Tamaulipas, located in warn-subhumid and semiwarm dry climates, respectively, both above the parallel 21°. During the autumn–winter season, only one locality reported more than 5000 ha, Tantoyuca, Veracruz, in a warm-subhumid climate. Four out of the 231 municipalities sowing corn grain were statistically significant for crop grain surface damages during the spring–summer season ($p < 0.05$), while five out of the 143 municipalities were significant for the autumn–winter season. These locations corresponded with the more extensive land surface used for agriculture. There is no statistical significance for surface loss and Disaster Declarations by drought during 2005, 2013, 2018, and 2019 in the autumn–winter season. Still, statistical significance exists between Disaster Declarations by drought for 2011 in the spring–summer season ($r = 0.425$, $p < 0.001$).

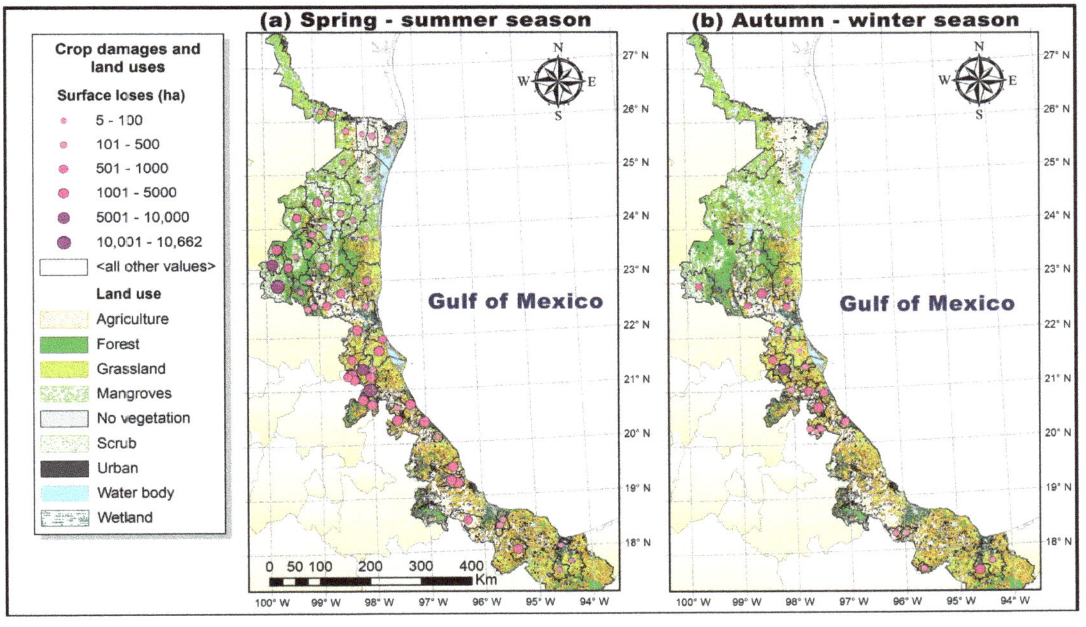


**Figure 12.** Damaged corn grain surfaces related to drought and land uses reported during 2003–2022 at the study site. (**a**) During spring–summer; (**b**) during autumn–winter. Source: land use and cartographic data were obtained from the latest update of the Mexican Institute of Geography and Statistical Information [28].

## 4. Discussion

### 4.1. Drought Levels in the Study Region

Since the Drought Monitor started recording drought data, drought has been registered in 91% of the municipalities during the spring–summer season and in 59% of the municipalities during the autumn–winter season. These results indicate that the study region is prone to drought, with the spring–summer season being the most sensitive to drought. The northern side of the area (above parallel 25°) has climates classified as warm-semidry and warm-dry [28]. Therefore, high average drought levels (D2, D3, and D4) should be considered as usual. Unfortunately, no functional meteorological stations were found at these latitudes to complement the analysis with the Drought Monitor. This situation emphasizes the need to consider the Drought Monitor as the only available tool to analyze the drought phenomena in the Mexican sites without available meteorological data.

In the south (below parallel 18°), three municipalities stand out for their high levels of average drought (D2), which should not be expected in a warm-humid climate, with average yearly precipitations between 2500 and 3500 mm [20]. They also do not have any available meteorological data [26]. Therefore, it is impossible to analyze these sites' climatic behavior. In this region, there has been a continued productive reconversion of the original tropical forests of the site to the current grasslands dedicated to cattle [20]. Therefore, these authors consider that the increment of the grasslands and the consequent decrease in the tropical forests is causing a climate change, transforming the place from a humid climate to a subhumid climate with lower precipitation levels, as has been observed in other sites of the region that have experienced similar transformations [30].

### 4.2. Disaster Declarations by Drought and Their Relationship with Corn Grain Production

Since the beginning of available national statistics at the municipal level, 2005, 2011, and 2019 were the years with the highest drought periods during Mexico's spring–summer and autumn–winter seasons [4]. At the same time, there were Drought Declarations in 2005, 2011, 2013, 2018, and 2019 in the study region, with 138 during the spring–summer and 167 during the autumn–winter season [25]. In this regard, the Disaster Declarations agree with the higher drought levels (D3 and D4) recorded for 2011, 2013, and 2019 for the municipalities registering corn grain losses due to drought during both seasons (spring–summer and autumn–winter). For 2005, the Mexican Drought Monitor reported lower drought levels (the D3 level was the highest). However, this year, 100% of the municipalities registered D0 or D1 levels below parallel 22.5° during January and April, 99% during February and May, and more than 57% during March and June [4]. Therefore, it can be considered that these three months affected by abnormal and moderate drought were catastrophic for corn grain production, increasing the damaged surface by 19% compared with the previous 2004 [11]. Complementing this information, the anomalies obtained with the climatological stations indicate that 2005 recorded higher temperatures than the average long-term data (Figure 6). Thus, these higher temperatures might have contributed to the damage caused by the droughts recorded in the 2005 national disaster records, especially during the autumn–winter season, when January registered temperatures above 75% of the long-term average, as seen in Figure 6. Higher temperatures and lower precipitations during December, January, April, and September (Figure 7) indicate that precipitation decrements affected both seasons, as shown in the Drought Monitor.

Unfortunately, the crops affected since 2011 were not recorded in the national agronomical statistics used in this research. The Mexican government replaced the direct economic support for agricultural damages with independently hired catastrophic insurance [25]. Under this new system, the insurance is only applied if there is 70% or more crop surface damage; otherwise, damaged croplands are not registered. This situation implies a subregister of data reported in the Mexican Agrifood Statistical System. For example, for 2019, the official records only indicate that corn was one of the crops affected by drought that received insurance payments. Still, they do not specify the names of the municipalities nor the surface receiving these insurance payments [31]. Table 4 shows a brief resume of data from different public sources regarding drought and its effects on corn production on the study site. As seen in the table, local data is scarce and incomplete. However, 2011 and

2019 are documented as years with total corn surface lost due to drought. This information confirms our study findings, which indicate that these municipalities reported corn grain surface damages during the Disaster Declarations due to drought.

**Table 4.** Documented drought and corn production were affected from 2003 to 2022 in the study site.

| Period | Site and Situation | Study Type and Reference |
| --- | --- | --- |
| 2019 | Atypical drought caused crop damage in 15 municipalities in Veracruz. A national increment in corn importation was documented due to drought. | State Agrifood Report [31]. Mexican newspaper [32]. |
| 2018–2020 | The municipality of El Mante, Tamaulipas, reported that 24% of its corn surface was loose due to a rain deficit. | Direct information from local producers [33]. |
| 2014 | The ten most productive municipalities in Veracruz reported decrements in corn yield due to drought and a deficit in government economic support. | Data was obtained from the National System of Agrifood Information and the state of Veracruz [34]. |
| 2011–2012 | The municipality of Tatahuicapan de Juárez, Veracruz, reported a total loss of corn production. | Data was obtained from the National System of Agrifood Information and the state of Veracruz [35]. |
| 2011 | San Fernando, Jaumave, Abasolo, González, Díaz Ordaz, Victoria, and El Mante, Tamaulipas, reported damaged corn lands due to drought during the spring–summer season. | Information was obtained from a Mexican newspaper [36]. |
| 2005 | Drought affected the northern side of Mexico and Veracruz, causing corn losses and affecting 950,000 ha of different crops. | Information was obtained from a Mexican newspaper [37]. |

### 4.3. *Drought Levels and Corn Grain Yield and Volume*

The number of significant correlations between yield and volume decrements and drought levels indicates that the spring–summer season has the most significant coefficient determinations (Figure 10). These results were obtained because more municipalities are sowing corn grain during this season; therefore, there were more substantial numbers of data (239 municipalities sown corn grain during the autumn-summer and only 151 during the autumn–winter season). For example, the droughts recorded during 2011 affected the spring–summer season more, with an average decrement of $-695$ kg ha$^{-1}$, compared to the autumn–winter season, which registered a decrement of $-8.57$ kg ha$^{-1}$. Low yield associated with drought conditions under rainfed corn has been previously documented for the eastern side of Mexico compared to irrigated corn production. However, there are no specific estimations of these decrements [38].

In Mexico, native corn varieties have been selected to be cultivated under the country's different agroclimatic conditions, especially considering drought tolerance and fast-growing and better performance. These studies have found that drought mainly affects chlorophyll contents, independently of the genetic origin, affecting the root and biomass development [39]. This situation implies that no matter how many new corn varieties are developed to tolerate drought, this plant will always be highly vulnerable. One study reported a drought-tolerant variety of white corn grain produced in the northern side of the study region. The authors indicated that this corn obtained an average harvest period of 100–110 days during the spring–summer season and 120–125 days during the autumn–winter season. Under experimental conditions and 350 mm of precipitation during the plant cycle, this variety reached a maximum yield of 3.78 tons·ha$^{-1}$ [40], above the average of the study site production of 2.44 tons·ha$^{-1}$. Nevertheless, the experimental data indicate that yield can decrease by 1.75 tons·ha$^{-1}$ under rainfed conditions compared with irrigated conditions in the same area.

Local research in the northern side of the region, under a warm subhumid climate, documented that the average rainfed corn grain yield is between 1.42 and 2.78 tons·ha$^{-1}$ [33]. This research also indicates that the local producers use certified and native corn grains adapted to their natural climatological conditions. However, these producers also manifest that drought is the leading cause of their low productivity and crop losses. Conversely, in the southern part of the study region, there are municipalities with averages between 3.86 and 4.01 tons·ha$^{-1}$ under a warm-humid climate [11]. These results indicate that precipitation is a crucial factor in incrementing yield. More rain and yield improvement have been demonstrated in specific studies where hydric stress played an important part when corn was flowering, decrementing its yield by 55%, even with a highly productive corn variety, when comparing a tropical humid forest versus a dry forest climate [41].

In a broad sense, national studies suggest that the minimal precipitation estimated for the Mexican corn was established at 480 mm during the plant's life cycle. Lower precipitations are not recommended under rainfed conditions [12]. Therefore, the results found in this study for the rainfed corn grain sown at the study site indicate that these national studies are more congruent with the corn varieties used in the study region. Studies like the one mentioned previously in the northern side of the study region and others in similar climate types [33,41] reporting yields between 3.4 and 3.7 tons·ha$^{-1}$ under drought conditions may need further review. The local statistics in this study indicate that producers may not apply the same care and techniques as the scientific reports document in their experiments. Additional studies document regional differences in production yields even with the same white corn variety. The regions with better yields have standard larger-scale production, improved germplasm, and intensive production systems. In comparison, the areas with lower yields have common environmental and substantial technological limitations [38].

Therefore, the study region may need more training in corn grain production and expert technical directions because national statistics indicate that 57% of these producers only finished elementary school (six years), and 14.8% do not have any education [42]. This situation makes it more difficult for corn cultivators to apply new technical developments and follow the instructions to use the new varieties developed by governmental research institutions for specific agroclimatic conditions [43].

The national statistics show that the average corn grain yield is 3.9 tons·ha$^{-1}$. However, this data includes rainfed and irrigated corn production [13]. For example, in the state with the highest corn grain production (Sinaloa on the northern and western sides of Mexico), irrigated corn grain has an average yield of 10.9 tons·ha$^{-1}$. In this same state, with high agricultural potential for corn, the rainfed corn averages 2.9 tons·ha$^{-1}$ [11], only slightly higher than the study region's average.

The other variable, corn grain volume (tons), is more related to the quantity of cultivated land during the period. Therefore, the municipalities with higher surfaces of corn sown during the driest seasons were the most affected and significantly correlated with drought levels reported by the Drought Monitor (Figures 8 and 9). Municipalities like Las Choapas, on the southern side of the region (with an average seasonal surface sown of 13,300 ha), registered −223.7 and −578.5 tons in the spring–summer season of 2011 and 2019 compared with their previous years. Papantla, in the region's center, with an average surface of 13,576 ha, recorded −3652 and −5966 tons in the same season of 2011 and 2019 [11].

Unfortunately, no meteorological stations were available for the Las Choapas and Papantla municipalities during the study period. Besides, a nearby Las Choapas site (Coatzacoalcos) did not contain rain precipitation data during 2011 but reported less than 30% of the normal rainfall from February to May 2019. Papantla's nearest meteorological station registered only 41% of the average rain precipitation from March to June and 80% of the normal from October to December 2011. During 2019, this station recorded 42% of the normal rainfall from July to September, affecting mainly the spring–summer season [26]. Therefore, we consider that, even with these scarce meteorological data, it is observable that corn grain was affected by low precipitations during 2011 and 2019. The year 2019 represented the most significant volume decrements because 2018 also had high drought levels in both municipalities (D0 from May to June, D1 and D2 for July, and D3 for August and September for Las Choapas, while D0 from June to August and D1 in September for Papantla). Thus, implementing more meteorological stations in the study site to correlate corn productivity and climate data is advisable to obtain better estimations.

Nevertheless, these results correlate with the higher volume decrements estimated with the linear regression equation obtained in this study. However, this equation underestimates the volume reduction for these two cases and obtains a low correlation with the drought levels registered at the sites. The lower correlation and coefficient of determination obtained with the linear models are congruent with sown fluctuations between seasonal periods. In this case, on average, only 2005 and 2019 recorded significantly decreased production volume due to drought. However, 2011 was not negatively affected in some municipalities of the study site compared with the previous year during the spring–summer season because 2010 registered one category-three hurricane that impacted the study region's center. However, in the significant municipalities, 2011 recorded 387.59 tons less than in 2010 during the spring–summer season and −269 tons during the autumn–winter season.

These peculiarities in corn production decrements and drought intensities indicate that other phenomena besides drought may affect corn grain production in the study region. This region is also affected by snow, hail, strong winds, excessive rains, floods, and tropical cyclones [25], which makes it more challenging to estimate drought and corn production correlations. Therefore, only the local meteorological stations can provide more reliable data to analyze the corn grain performance related to drought levels.

## 5. Conclusions

The North American Drought Monitor and its Mexican part, the Mexican Drought Monitor, can be a helpful tool to predict corn grain productivity behavior in specific sites without local meteorological data and when other phenomena, like tropical cyclones, floods, or excessive rains, were not registered. Under these situations, drought intensities with levels D3 (extreme drought) and D4 (exceptional drought) during the spring–summer season and autumn–winter season may cause a total loss of corn grain production, mainly in warm-subhumid and temperate-subhumid climates. During the spring–summer season, crop surface damage can be 4.7 times higher than during the autumn–winter season due to the more considerable land extension cultivated during this season. During the spring–summer season, corn grain yield can decrease by 1.77 tons·ha$^{-1}$ and 1.38 tons·ha$^{-1}$ during the autumn–winter season with D2 and D3 drought levels. The spring–summer season has a higher predictability than the autumn–winter season due to the higher number of municipalities and land surfaces cultivating this crop during this period, which provides a more considerable amount of data to correlate drought and corn grain production.

We suggest more detailed in situ studies of drought levels and corn grain production, considering corn varieties and cultivation techniques, to improve the predictability of the drought effects on corn grain production under rainfed conditions.

**Author Contributions:** Conceptualization, O.A.V.-R.; methodology, O.A.V.-R.; software O.A.V.-R.; validation, O.P.-W., F.S.-M. and A.M.; formal analysis O.A.V.-R. and F.S.-M.; investigation O.A.V.-R. and F.S.-M.; resources, O.P.-W. and A.M.; data curation, O.A.V.-R.; writing—original draft preparation,

O.A.V.-R.; writing—review and editing, O.P.-W.; visualization, A.M. All authors have read and agreed to the published version of the manuscript.

**Funding:** This research was funded by the grant supplied by the CONAHCYT to the Mexican researchers.

**Institutional Review Board Statement:** Not applicable.

**Informed Consent Statement:** Not applicable.

**Data Availability Statement:** Meteorological data can be obtained by the National Meteorological Service at https://smn.conagua.gob.mx/es/climatologia/informacion-climatologica/informacion-estadistica-climatologica. Accessed 5 February 2024. The Drought Monitor data can be consulted at: https://smn.conagua.gob.mx/es/climatologia/monitor-de-sequia/monitor-de-sequia-en-mexico Accessed 5 February 2024. The agrifood data can be consulted at: http://infosiap.siap.gob.mx/gobmx/datosAbiertos.php Accessed 5 February 2024.

**Acknowledgments:** To the CONAHCYT for their funding support.

**Conflicts of Interest:** The authors declare no conflicts of interest.

## Appendix A

**Table A1.** Geographical location of the meteorological stations used in this study.

| Meteorological Station | Municipality/State | Latitude (°) | Longitude (°) | Altitude (msnm) |
|---|---|---|---|---|
| Juan Rodríguez Clara | Juan Rodríguez Clara, VER. | 17.9930556 | −95.402778 | 148 |
| El Raudal | Nautla, VER. | 20.1561111 | −96.721111 | 10 |
| Tecolutla | Tecolutla, VER. | 20.4791667 | −97.009722 | 7 |
| Actopan | Actopan, VER. | 19.5027778 | −96.611111 | 250 |
| Alamo | Álamo Temapache, VER. | 20.9294444 | −97.679444 | 19 |
| Angel R. Cabada | Ángel R. Cabada, VER. | 18.5972222 | −95.447222 | 28 |
| Villa Tejeda | Camarón de Tejeda, VER. | 19.0222222 | −96.613889 | 348 |
| Chicontepec De Tejeda (SMN) | Chicontepec, VER. | 20.9933333 | −98.163889 | 291 |
| Coatzacoalcos (OBS) | Coatzacoalcos, VER. | 18.1402778 | −94.522222 | 16 |
| Rancho Viejo | Emiliano Zapata, VER. | 19.4469444 | −96.783611 | 914 |
| El Tejar | Medellín de Bravo, VER. | 19.0672222 | −96.158333 | 10 |
| Pánuco (DGE) | Pánuco, VER. | 22.0591667 | −98.175556 | 11 |
| Loma fina | Paso de Ovejas, VER. | 19.2613889 | −96.41 | 41 |
| Tempoal de Sánchez | Tempoal, VER. | 21.5188889 | −98.410278 | 34 |
| Tuxpan (OBS) | Tuxpan de Rodríguez Cano, VER. | 20.9597222 | −97.418889 | 5 |
| José Cardel | La Antigua, VER. | 19.3647222 | −96.374444 | 28 |
| Platón Sánchez | Platón Sánchez, VER. | 21.2980556 | −98.356389 | 57 |
| Poza Rica | Poza Rica de Hidalgo, VER. | 20.5408333 | −97.472778 | 50 |
| San Juan Evangelista (DGE) | San Juan Evangelista, VER. | 17.8833333 | −95.145833 | 18 |
| Ciudad Victoria (OBS) | Victoria, TAM. | 23.7477778 | −99.171667 | 336 |
| Soto la Marina (OBS) | Soto la Marina, TAM. | 23.7666667 | −98.2 | 21 |
| San Nicolás | San Nicolás, TAM. | 24.6894444 | −98.829722 | 797 |
| Palmillas | Palmillas, TAM. | 23.3022222 | −99.548333 | 1264 |
| Nuevo Morelos | Nuevo Morelos, TAM. | 22.5283333 | −99.213889 | 260 |

**Table A1.** *Cont.*

| Meteorological Station | Municipality/State | Latitude (°) | Longitude (°) | Altitude (msnm) |
|---|---|---|---|---|
| Miquihuana | Miquihuana, TAM. | 23.5738889 | −99.753056 | 1851 |
| E.T.A. 067 CIUDAD MANTE | El Mante, TAM. | 22.7425 | −98.972222 | 91 |
| Güémez | Güémez, TAM. | 23.9186111 | −99.004444 | 229 |
| La Mayeb | González, TAM. | 22.9155556 | −98.345556 | 1371 |
| Ahualulco | Gómez Farías, TAM. | 22.9886111 | −99.145 | 99 |
| Cruillas | Cruillas, TAM. | 24.7541667 | −98.534722 | 229 |
| Paso de Molina | Casas, TAM. | 23.725 | −98.745278 | 150 |
| Bustamante | Bustamante, TAM. | 23.4358333 | −99.754167 | 1666 |
| El Carrizal | Aldama, TAM. | 22.8586111 | −98.233333 | 90 |
| San Gabriel | Xicoténcatl, TAM. | 23.0841667 | −98.7875 | 135 |
| Conrado Castillo | Villagrán, TAM. | 24.6669444 | −99.253889 | 309 |
| Jaumave (DGE) | Jaumave, TAM. | 23.4075 | −99.375278 | 324 |
| El Barranco | Altamira, TAM. | 22.5658333 | −97.905 | 6 |
| Abasolo (DGE) | Abasolo, TAM. | 24.0655556 | −98.39 | 70 |

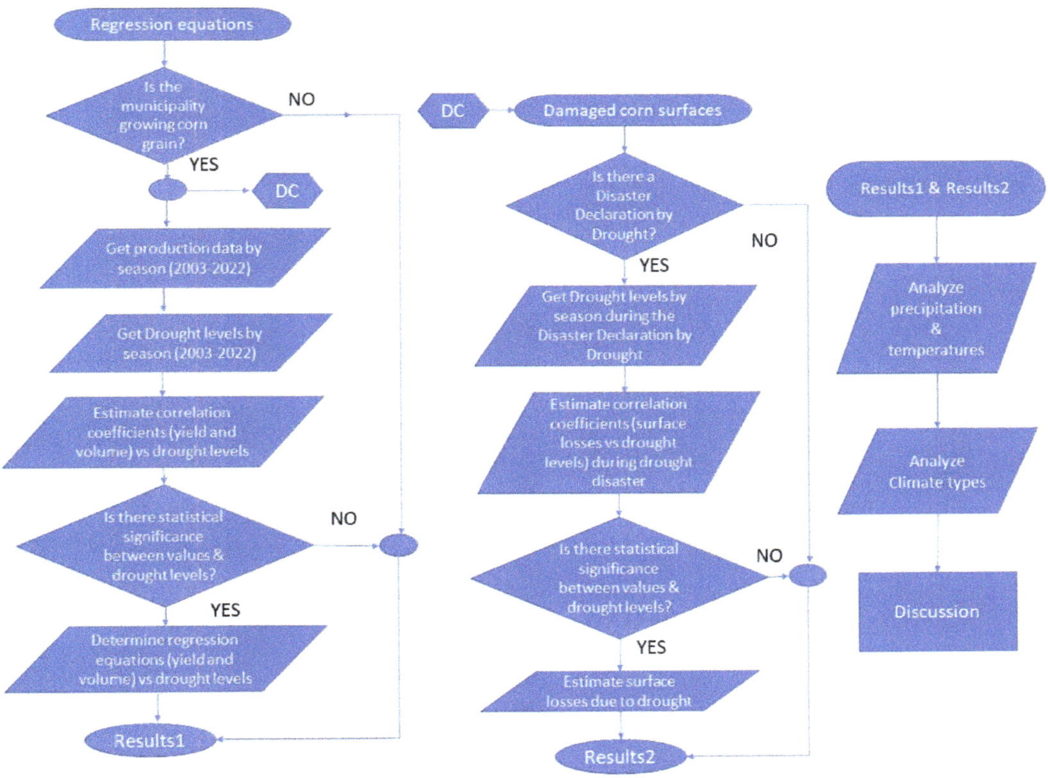


**Figure A1.** Analysis process to determine the relationship between drought levels and corn grain productivity and between drought levels and damaged corn grain surfaces.

## References


1. NDMC, (National Drought Mitigation Center) Drought Basics. Available online: https://drought.unl.edu/Education/DroughtBasics.aspx (accessed on 17 May 2024).
2. García Jimenez, F.; Fuentes Mariles, O.; Matías Ramírez, L.G. *Fascículo Sequías*, 2nd ed.; Ramos Padilla, V., Ed.; Secretaría de Seguridad y Protección Ciudadana: Mexico City, Mexico, 2014; ISBN 9706286136.
3. Hadwen, T.; Lopez, M.; Pascual, R.; Pugh, B.; Simeral, D. North American Drought Monitor. Available online: https://droughtmonitor.unl.edu/nadm/Home.aspx (accessed on 18 December 2024).
4. SMN, (Servicio Meteorológico Nacional) Monitor de Sequía En México. Available online: https://smn.conagua.gob.mx/es/climatologia/monitor-de-sequia/monitor-de-sequia-en-mexico (accessed on 12 June 2023).
5. Leeper, R.D.; Bilotta, R.; Petersen, B.; Stiles, C.J.; Heim, R.; Fuchs, B.; Prat, O.P.; Palecki, M.; Ansari, S. Characterizing U.S. Drought over the Past 20 Years Using the U.S. Drought Monitor. *Int. J. Climatol.* **2022**, *42*, 6616–6630. [CrossRef]
6. SMN, (Servicio Meteorológico Nacional) Índice Estandarizado de Precipitación (SPI). Available online: https://smn.conagua.gob.mx/es/climatologia/monitor-de-sequia/spi (accessed on 25 October 2024).
7. WMO, (World Meteorological Organization). *Estado Del Clima En América Latina y El Caribe 2023*; Presidente de la Junta de Publicaciones, Ed.; CMM-N° 135.; Organización Meteorológica Mundial: Ginebra, Suiza, 2024; ISBN 978-92-63-31322-5.
8. Domínguez, J. Revisión Histórica de Las Sequías En México: De La Explicación Divina a La Incorporación de La Ciencia. *Tecnol. y Ciencias del Agua* **2016**, *7*, 77–93.
9. García Acosta, V.; Padilla Lozoya, R. *Historia y Memoria de Los Huracanes y Otros Episodios Hidrometeorológicos Extremos En México*, 1st ed.; García Acosta, V., Padilla Lozoya, R., Eds.; Universidad de Colima, Universidad Veracruzana, Centro de Investigaciones: Colima, Mexico, 2021; Volume 5, ISBN 9781119130536. Available online: https://libros.uv.mx/index.php/UV/catalog/book/FC304 (accessed on 19 December 2024).
10. SIAP, (Servicio de Información Agroalimentaria y Pesquera). *Panorama Agroalimentario 2022*; SAGARPA, (Secretaría de Agricultura y Desarrollo Rural): México City, Mexico, 2022.
11. SIAP, (Servicio de Información Agroalimentaria y Pesquera) Estadística de Producción Agricola. Available online: http://infosiap.siap.gob.mx/gobmx/datosAbiertos.php (accessed on 30 September 2024).
12. SIAP-SADER. *Aptitud Agroclimática Del Maíz En México*, 1st ed.; Geoespacireales, D.d.S., Ed.; Secretaría de Agricultura y Desarrollo Rural: Mexico City, Mexico, 2018; Volume 1.
13. SIAP, (Servicio de Información Agroalimentaria y Pesquera). *Panorama Agroalimentario 2023*, 2023rd ed.; Secretaría de Agricultura y Desarrollo Rural: Mexico City, México, 2023; Volume 1.
14. Valdés-Rodríguez, O.A.; Salas-Martínez, F.; Palacios-Wassenaar, O.M. Hydrometeorological Hazards on Crop Production in the State of Veracruz, Mexico. *Atmosphere* **2023**, *14*, 287. [CrossRef]
15. Gutiérrez Núñes, N.L. La Agricultura Del Maíz y El Sorgo En El Bajío Mexicano: Revolución Verde, Sequías y Expansión Forrajera, 1940-2021. *Hist. Agrar.* **2023**, 255–286. [CrossRef]
16. Salas Martínez, F.; Valdés Rodríguez, O.A.; Palacios Wassenar, O. Reducciones Sobre El Rendimiento Del Maíz En Grano En La Región Central de Veracruz, México. In *Ciencia y Tecnología Para el Campo Mexicano: Retos y Oportunidades*; Zetina Lezama, R., Tosquy Valle, O.H., del Angel Pérez, A.L., Ríos Utrera, Á., Esqueda Esquivel, V.A., Eds.; INIFAP: Mexico City, Mexico, 2021; pp. 1530–1541. ISBN 978-607-37-1396-2.
17. Cruz-González, A.; Arteaga-Ramírez, R.; Sánchez-Cohen, I.; Monterroso-Rivas, A.I. Impacts of Climate Change on Corn Production in Mexico. *Rev. Mex. Ciencias Agrícolas* **2024**, *15*, 1–13. [CrossRef]
18. SMN, (Servicio Meteorológico Nacional) Funciones y Objetivos Del SMN. Available online: https://smn.conagua.gob.mx/es/smn/funciones-y-objetivos (accessed on 2 May 2024).
19. Luna Díaz Peón, A. *Análisis de La Red Climatológica Del Estado de Veracruz y Propuestas de Estimación de Precipitación Para Eventos Por Riesgos Hidrometeorológicos*; El Colegio de Veracruz: Xalapa, Mexico, 2019.
20. INEGI, (Instituto Nacional de Estadística y Geografía). *Aspectos Geográficos Veracruz*; INEGI: Mexico City, Mexico, 2021; Available online: https://www.inegi.org.mx/contenidos/app/areasgeograficas/resumen/resumen_30.pdf (accessed on 19 December 2024).
21. INEGI, (Instituto Nacional de Estadística y Geografía). *Aspectos Geográficos Tamaulipas*; INEGI: Mexico City, Mexico, 2021; ISBN 7028251977.
22. Salas-Martínez, F.; Márquez-Grajales, A.; Valdés-Rodríguez, O.-A.; Palacios-Wassenaar, O.-M.; Pérez-Castro, N. Prediction of Agricultural Drought Behavior Using the Long Short-Term Memory Network (LSTM) in the Central Area of the Gulf of Mexico. *Theor. Appl. Climatol.* **2024**, 1–21. [CrossRef]
23. Wagner, R.; Grimm, M.S. Empirical Validation of the 10-Times Rule for SEM. *Springer Proc. Bus. Econ.* **2023**, 3–7. [CrossRef]
24. USDA, U.S.D. of A. Row Crops. Available online: https://agindrought.unl.edu/RowCrops.aspx (accessed on 19 December 2024).
25. CENAPRED, (Centro Nacional de Prevención de Desastres) Sistema de Consulta de Declaratorias. Available online: http://www.atlasnacionalderiesgos.gob.mx/apps/Declaratorias/ (accessed on 22 August 2023).

26. SMN, (Servicio Meteorológico Nacional) Información Estadística Climatológica. Available online: https://smn.conagua.gob.mx/es/climatologia/informacion-climatologica/informacion-estadistica-climatologica (accessed on 30 October 2024).
27. OMM, (Organización Meteorológica Mundial). *Guía de Prácticas Climatológicas*, 100th ed.; Mundial, O.M., Ed.; Organización Meteorológica Mundial: Geneva, Switzerland, 2018; ISBN 9789263301000.
28. INEGI, (Instituto Nacional de Estadística y Geografía) Mapa Digital. Available online: https://www.inegi.org.mx/temas/mg/ (accessed on 19 September 2024).
29. INEGI, (Instituto Nacional de Estadística y Geografía) Mapa Topográfico de Veracruz de Ignacio de La Llave. Available online: https://www.inegi.org.mx/app/mapas/ (accessed on 10 January 2024).
30. Barradas, V.L.; Cervantes Pérez, J.; Anyul, C.P. *Evidencia de Un Cambio Climático En La Región de Las Grandes Montañas Del Estado de Veracruz, México*; Asociación Española de Climatología: Santander, UK, 2004.
31. Santos Ramos, Y.; Sánchez Román, J. *Seguro Agrícola Catastrófico*; SEDARPA: Xalapa-Enríquez, Mexico, 2024.
32. González, G.S. México Importará Cifra Récord de Maíz Este Año Por La Sequía. Available online: https://www.jornada.com.mx/noticia/2020/01/21/economia/mexico-importara-cifra-record-de-maiz-este-ano-por-la-sequia-8730 (accessed on 8 June 2021).
33. Garay Martínez, J.R.; Barrón Bravo, O.G.; Granados Rivera, L.D.; Patricia, M.T.S.; Avilés Ruiz, R. Caracterización de Las Unidades de Producción de Maíz y Sorgo En El Mante, Tamaulipas. *Transversalidad Científica y Tecnológica* **2020**, *4*. Available online: https://www.researchgate.net/profile/Jesus-Reyna-Fuentes/publication/349949728_SITUACION_ACTUAL_DE_LAS_EMPRESAS_APICOLAS_EN_TAMAULIPAS_MEXICO/links/6048de0d45851543166b5d12/SITUACION-ACTUAL-DE-LAS-EMPRESAS-APICOLAS-EN-TAMAULIPAS-MEXICO.pdf#page=49 (accessed on 19 December 2024).
34. Bada-Carbajal, L.M.; Osorio-Antonia, J.; Ramírez-Hernández, Z. Evolución de La Producción Del Maíz En Veracruz, México. *Investig. Adm.* **2021**, *50–2*, 1–16. [CrossRef]
35. Guerra Martínez, A.; López Galindo, L.K.I.; Álvarez Ramírez, M.M.; Antonio Sánchez, D.G. Caracterización de La Sequía En El Estado de Veracruz (2007-2018) y Su Efecto En La Seguridad Alimentaria. *Obs. Segur. Aliment. y Nutr. Veracruz* **2020**, 272–284.
36. Sin Embargo En Tamaulipas Activan Seguro Agrícola Catastrófico. *SinEmbargo MX*. 2011. Available online: https://www.sinembargo.mx/61939/en-tamaulipas-activan-seguro-agricola-catastrofico/ (accessed on 19 December 2024).
37. Balboa, J. Sequía Atípica En Seis Estados Afectará Cultivos y Casi 3 Millones de Animales. *La Jorn.* 2005. Available online: https://www.jornada.com.mx/2005/11/27/index.php?section=sociedad&article=040n1soc (accessed on 19 December 2024).
38. González-Cruz, J.L.; Torres-Rojo, J.M. Regional Differences in Rainfed White Corn Production in Mexico. *Rev. Mex. Ciencias Agrícolas* **2024**, *15*, e3170. [CrossRef]
39. Félix-Lizárraga, J.U.; Ruiz-Torres, N.A.; Rincón-Sánchez, F.; Sánchez-Ramírez, F.J.; Borrego-Escalante, F.; Benavides Mendoza, A. Selección de Poblaciones de Maíz Con Base En La Producción Temprana de Biomasa Bajo Condiciones de Estrés Salino. *Rev. Mex. Ciencias Agrícolas* **2023**, *14*, 449–458. [CrossRef]
40. Augusto Reyes Méndez, C.; Ángel Cantú Almaguer, M.; Vázquez Carrillo, G. H-440, Nuevo Híbrido de Maíz Tolerante a Sequía Para El Noreste de México. *Agric. Técnica México* **2007**, *33*, 201–204.
41. Díaz-Chuquizuta, P.; Hidalgo-Melendez, E.; Mendoza-Paredes, M.; Cieza-Ruiz, I.; Jara-Calvo, T.W.; Valdés-Rodríguez, O.A. Nuevo Híbrido Trilineal de Maíz Amarillo Duro Para El Trópico Peruano. *Agron. Mesoam.* **2022**, *34*, 51177. [CrossRef]
42. INEGI, (Instituto Nacional de Estadística y Geografía) El Inegi y Agricultura Presentan Los Resultados de La Encuesta Nacional Agropecuaria (ENA). 2019. Available online: https://www.gob.mx/agricultura/prensa/el-inegi-y-agricultura-presentan-los-resultados-de-la-encuesta-nacional-agropecuaria-ena-2019?idiom=es#:~:text=InegiyAgriculturadana, produceenelcampomexicano (accessed on 2 January 2022).
43. INIFAP, (Instituto Nacional de Investigaciones Agrícolas y Pecuarias). *Agenda Técnica Agrícola de Veracruz*, 1st ed.; SAGARPA: Mexico City, Mexico, 2017.

MDPI AG
Grosspeteranlage 5
4052 Basel
Switzerland
Tel.: +41 61 683 77 34

*Atmosphere* Editorial Office
E-mail: atmosphere@mdpi.com
www.mdpi.com/journal/atmosphere

www.ingramcontent.com/pod-product-compliance
Lightning Source LLC
Chambersburg PA
CBHW061626210926
51850CB00009B/2450
* 9 7 8 3 7 2 5 8 5 5 0 1 8 *